Poultry Production Systems

Behaviour, Management and Welfare

Poultry Production Systems

Behaviour, Management and Welfare

Michael C. Appleby
Institute of Ecology and Resource Management
University of Edinburgh
School of Agriculture Building
West Mains Road
Edinburgh EH9 3JG, UK

Barry O. Hughes
AFRC Institute of Animal Physiology and Genetics Research
Edinburgh Research Station
Roslin
Midlothian EH25 9PS, UK

H. Arnold Elson
Agricultural Development and Advisory Service
Block 7, Chalfont Drive
Nottingham NG8 3SN, UK

C·A·B International

C·A·B International   Tel: Wallingford (0491) 32111
Wallingford   Telex: 847964 (COMAGG G)
Oxon OX10 8DE   Telecom Gold/Dialcom: 84: CAU001
UK   Fax: (0491) 33508

A catalogue record for this book is available from the British Library

ISBN 0 85198 797 4

Typeset by Hope Services (Abingdon) Ltd
Printed and bound in Great Britain by
Redwood Press Ltd, Melksham

CONTENTS

Foreword

by Professor Paul B. Siegel

This book takes an approach quite different from that usually seen in books involving management and husbandry of livestock and poultry. Statistics for production worldwide or for specific geographic areas are not presented, nor is there information on breeder improvement programmes, infectious diseases or nutrition. Rather, emphasis is on husbandry systems and their interface with behaviour. The interface is described as to how poultry adapt to husbandry practices and how husbandry should relate to behaviours. A common thread throughout the book is the welfare of the individual animal with major emphasis on chickens.

Because of its emphasis on behaviour, this book fills a void that exists in most books on poultry production. Overviews are provided on the special senses and the nervous, reproductive, digestive, skeletal and respiratory systems as they enhance an understanding of the biology of behaviour. A series of 'bullets' used to highlight key points at the beginning of each chapter is followed in most cases by a brief background on jungle and/or feral fowl as they relate to the topic which will be presented. The book is not highly technical and there are numerous figures which enhance readability for a general, as well as a more specialized, audience. Heavy reliance on literature from Western Europe provides insights to those in other areas of the world who are not familiar with that literature.

Concern for animal welfare in poultry production is growing in America and behavioural aspects are part of the orchestration. Appleby, Hughes and Elson describe alternative husbandry systems such as aviaries, percheries and free range, as well as criteria for these systems as outlined by the European Community. The economics of alternative systems is presented not only from the viewpoint of costs of production, but also with regard to the effects of legislation on viability of units in a multinational economy.

Husbandry systems are discussed not only as they influence the welfare of the birds but also as they influence the welfare of those persons who are their caretakers. The conundrum of profitability of the individual animal versus that of the flock is addressed, as are aspects of public opinion and legislation on welfare and husbandry practices. They point out that in legislation there is a difference between defining how animals should be kept (prescriptive) rather than how they should not be kept (proscriptive).

Poultry Production Systems: Behaviour, Management and Welfare fills a niche in providing readers with insights from the Western European experience. As such it is important reading for those in other parts of the world where poultry is, or is becoming, a major agricultural enterprise.

Professor Paul B. Siegel
Virginia Polytechnic Institute and State University
Blacksburg, Virginia, USA

Foreword

by Professor Rose-Marie Wegner

With this book a new approach is taken up by the three well-known authors who, for the first time, are combining basic and applied details on poultry production with poultry behaviour and poultry welfare. I am convinced that most readers will agree with me that this combination is not only a new and interesting one but also very commendable. When the extensive discussions about poultry production systems and welfare started about fifteen years ago, I remember very well that there was not always a very good consensus between the scientists working in the different disciplines of poultry behaviour and poultry management – most regrettably! Therefore it is especially noteworthy that this book is written by two behaviour scientists in cooperation with an international consultant, an expert in poultry systems, who has worked for many years with ADAS, the Agricultural Development and Advisory Service of the British Ministry of Agriculture, Fisheries and Food.

Another innovation is the summary at the beginning of each chapter, in which the most important conclusions or principles are shown. They make the reader curious to know more about the details and might help hasty readers to find a special subject that they are searching for.

It will be seen that up to now most relevant research has been on fowl, with relatively little on other poultry species. Although most of the experiments are also relevant to other poultry species, scientists may be encouraged to involve other species of poultry in similar experiments in the future. A lot of research is still needed on poultry behaviour in connection with poultry systems.

I myself have been involved in discussions concerning the welfare of laying hens in cages since 1970 and in Germany I made extensive investigations into the development of alternative systems, starting in 1976 at Bonn University and continuing during my directorship of the Federal Institute of Poultry Research at

Celle until my retirement in 1989. In the late 1970s cooperation in that field began with many other scientists in several European countries, partly initiated by the European Community (EC) in Brussels and partly by Working Group IX of the European Federation of Branches of the World's Poultry Science Association. This was continued by the first and second Symposia on Poultry Welfare in Denmark in 1981 and in Celle in 1985. During the last 10 years I had the privilege of being associated with the authors of this book in different experimental work that we have undertaken together, and I was always impressed by the very good cooperation and the fruitful discussions we had together.

Recently the involvement of British scientists and research stations has increased remarkably. Therefore it is not surprising that all three authors of this book are British. It seems to be almost a tradition: the first, and to my knowledge the only specialist book on hen battery systems was also composed by a British scientist: *Hen Batteries*, by Dr W. P. Blount (1951). I was so fascinated by this book that I translated it into German, but it was never published in German.

I am convinced that this new book will fill a gap – it will contribute to a better understanding between the disciplines of poultry management, poultry behaviour and poultry welfare and it will be useful for students, advisers and scientists involved in poultry production, behaviour and welfare. The book will be of benefit to readers in many countries, not only in Europe but also in other continents, because the importance of poultry welfare is still increasing around the world. It is therefore of special value that it is published simultaneously in the UK and in America.

About forty years ago I found in a British book on poultry keeping a remarkable quotation of Francis Bacon from about 400 years ago:

> Some books are to be read only in parts; others to be read, but not curiously; and some few to be read wholly and with diligence and attention.

I am convinced that this book is worthy to be placed in the third category!

Professor Rose-Marie Wegner
Celle, Germany

Preface

In this book we set out to review the poultry production industry and the husbandry systems which are available or now being developed. In doing this, we can draw on extensive strategic and applied research on poultry biology and behaviour which is highly relevant to commercial production. Research on poultry systems is an active and developing subject, as can be seen from the high proportion of the references cited which have been published within the last ten years. Many of the key references, in fact, have been published within the last three years.

The book considers the ways in which the biology and behaviour of the birds concerned influence the performance of different systems and are in turn influenced by the environment. It is evident that poultry biology is fundamental to the process of production and should be taken fully into account in the design and management of systems. In Part II, husbandry systems are covered in a cross-sectional way, by examining how birds behave in different systems and considering the implications for management and welfare. This part covers key aspects of behaviour and their interaction with the environment, opportunities for using them to best advantage and problems which may arise.

The approach of the book can perhaps be conveyed by the question 'which came first, the chicken or the poultry farm?' Poultry production is basically a biological process, not a technical one. As such, a sound understanding of poultry behaviour and other aspects of biology is an important prerequisite for designing and maintaining husbandry systems and for preventing or overcoming management problems. Systems designed from a biological perspective will have fundamentally better prospects for the welfare of their stock, will be more satisfactory for the producer and will be more acceptable to the public. However, this approach may also have implications for the economic performance of systems. Throughout the book we attempt to achieve a balance between the needs of the birds and the needs of a producer who has to make a profit to stay in business. In doing this, we hope that we have summarized the results of an extremely active area of scientific research in a way that is useful for those involved in the field, while also making them accessible to the general reader.

Our own involvement with the issue of poultry welfare stems partly from a concern for the birds themselves, partly from a practical interest in the poultry industry and the costs and benefits which can accrue from taking welfare into account, and partly from a belief that change in the industry, for example change imposed by legislation, should have a sound scientific basis. All these concerns are integral to the book and this diversity of interest has implications for our approach. In particular, it would be a mistake to assume that all our readers will have similar views on welfare. There is, however, a general consensus that animal welfare does exist: to put it another way, most people believe that animals can suffer in ways that are comparable to human suffering. Welfare is not a simple scale, though, from bad to good. It has many different aspects, in the same way as there are different kinds of suffering. For example, the UK's Farm Animal Welfare Council has proposed (Webster and Nicol, 1988) that a husbandry system should provide animals with: freedom from hunger and thirst; freedom from thermal and physical discomfort; freedom from pain, injury and disease; freedom from fear and distress; freedom to exercise most normal patterns of behaviour. Some people regard this as a realistic proposal and some as an idealistic one. We quote it here as a useful expression of the complexity of animal welfare.

We are indebted to friends and colleagues for assistance and discussions during the writing of this book. In particular, Ben Mather made invaluable recommendations on the whole book and the following all read sections or chapters and contributed useful suggestions: Melanie Adcock, Werner Bessei, Harry Blokhuis, Bjarne Braastad, Jean-Michel Faure, Detlef Fölsch, Mike Gentle, Martina Gerken, Bill Jackson, Fris Jensen, Jos Metz, Andrew Mills, Carol Petherick, Mark Rutter, John Savory, Ragnar Tauson, Klaus Vestergaard, Piet Wiepkema and David Wood-Gush. Elliot Armstrong was responsible for drawing or redrawing many of the figures; Roddy Field and Norrie Russell helped us with a number of the photographs. We are grateful to EB Equipment Ltd for permission to reproduce a photograph of a pan feeder. Grace Owens collated the text from several word processors with good nature and efficiency. We are very grateful to Professors Paul Siegel and Rose-Marie Wegner for their generous Forewords and to Sharon Cooper and the staff of Butterworth-Heinemann and also to CAB International for their professional guidance. The most important support, however, came from our wives, and to them we unreservedly dedicate this book.

Michael Appleby,
Barry Hughes
and Arnold Elson

To

Elaine Appleby, Helen Hughes and Kathy Elson

I

Poultry Production

Part I presents a review of the poultry production industry and of
the main factors which influence it. Principal among these must be
the biology of the birds concerned, and of the ways in which they
interact with the environment: the implications of these for the
structure of the industry and of husbandry systems are considered.
The influence of the wider issues of economics and legislation is
then examined.

1

Origins and biology

1.1 Summary

● All domestic poultry come from three Orders, the Galliformes (including fowl), the Anseriformes (ducks and geese) and the Columbiformes (pigeons). This emphasizes the fact that their progenitors must have shared biological features, including aspects of behaviour, which predisposed them for domestication.

● Domestic fowl were domesticated from Red Jungle Fowl over 8000 years ago. Early selection was for behaviour (fighting) and other purposes as well as production, and the basis of modern breeds existed by Roman times. Other poultry have also been selected over many centuries, more exclusively for eggs and meat. Nevertheless, most behaviour shown by wild relatives is also seen in modern forms.

● The avian central nervous system is well developed, especially in areas concerned with control of flight. Visual development is also partly associated with flight and partly with complex social behaviour. Other senses are well developed, too, including sensitivity to touch and pain.

● Skeletal, muscular and respiratory systems are adapted to flight. For purposes of production this results in both advantages (such as the large breast muscles) and disadvantages (such as the aerated bones which are strong but liable to break on impact).

● The reproductive and digestive systems are those most affected by selection and by environmental conditions. Laying strains of domestic fowl now produce over 300 eggs in 365 days. Both egg-laying and meat strains have been selected for efficiency: maximum output for minimum food intake.

● The considerable knowledge now available on the biology of poultry can be used in housing design, especially by identifying those features which can be modified and controlled, and those which need to be accommodated.

1.2   Origins of domestic fowl

The progenitor of the domestic fowl was the Red Jungle Fowl (*Gallus gallus*), modern forms of which are found in Central and South India (*Gallus gallus sonnerati*), East India (*G. g. murghi*), Burma and Malaysia (*G. g. spadiceus*) and Thailand and Cambodia (*G. g. gallus*). It is a smaller bird than most domestic varieties – an adult female weighs about 800 g – and it is a tropical species. Along the Himalayan foothills its range is bounded by the 10°C isotherm, and it is typically confined to forested areas and to thick vegetation. However, the jungle fowl's ability to adapt to a broad range of environments, together with its potential genetic variability, subsequently helped the domestic fowl to become widely distributed throughout the world.

It is believed that the fowl was first domesticated in Southeast Asia, probably well over 8000 years ago (Yamada, 1988). The Latin nomenclature of the domestic fowl has been a source of controversy, but it is now accepted that it is not a separate species. It should, instead, be regarded as a sub-species of the jungle fowl and thus should be called *Gallus g. domesticus*. Until recently it was thought that the fowl reached Europe via India and the Middle East, because of the evidence of domestication in the Indus Valley around 2000 BC (Sewell and Guha, 1931; Zeuner, 1963) and the fact that it was known to be present in the area around Babylon in 2400 BC and in Egypt in 1400 BC. Early Sumerian texts contain the word for cock and there is a clear representation of a galliform cockerel outside Tutankamun's tomb (Coltherd, 1966). However, recent archaeological research, based primarily on the existence of chicken bones in deposits of known age, has shown that after domestication they were first taken north (Figure 1.1); domestic fowls were established in China by 6000 BC (West and Zhou, 1989). From there it is believed that they were taken across the Russian steppes and there is evidence of their presence in Turkey and in Eastern Europe (Romania, Greece) during the later Stone Age (3000 BC). By 1200 BC they had reached Spain, and they were to be found in Northwest Europe by about 500 BC (West and Zhou, 1989). The earliest records in Britain date from 100 BC (Brown, 1929) and they were introduced to North America about 1550 AD (Yamada, 1988).

During the earlier stages of domestication, the fowl was probably valued mainly as a sacrificial or religious bird, or for cockfighting. It was the Romans who developed its potential as an agricultural animal, creating specialized breeds (Thomson, 1964), including very productive layers, and forming a complex poultry industry, which paid close attention to rearing, housing, disease control, costing and marketing (Wood-Gush, 1959a). Pliny wrote that in Roman times there were birds laying an egg every day (Wood-Gush, 1971). With the

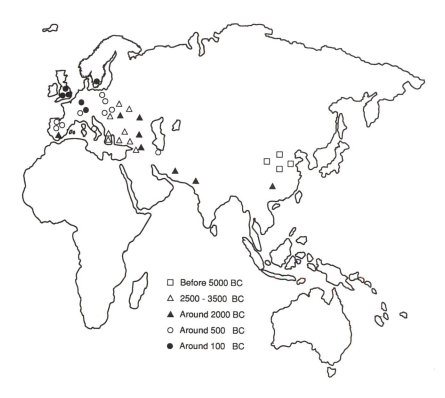

Figure 1.1.   Archaeological evidence shows that the fowl was first domesticated in Southeast Asia and subsequently spread north and west through China. The symbols show the earliest dates of evidence of domesticated fowls at the various locations (after West and Zhou, 1989).

decline of the Roman Empire the industry collapsed and it did not resume on a large scale until the nineteenth century.

1.3   Breeds of domestic fowl

Wood-Gush (1959a) identifies several breeds in Roman times – two heavy fighting breeds, two dual-purpose ones, native Roman breeds and an especially prolific variety from Adria. Subsequently, little systematic selection was practised for many centuries, with the exception of birds for cockfighting. The appreciable levels of bird-to-bird aggression, which can pose a problem in some modern poultry-keeping systems, may be a legacy of this approach.

It was only in the nineteenth century that the situation changed, with an explosion of poultry breeding. Six breeds are mentioned as existing in England around 1810: the Game, the White or English, the Black or Poland, the

Darkling, the Large or Strakeberg and the Malay (Wood-Gush, 1959a). In the next 50 years formal poultry shows were organized, many novel breeds were created and numerous Breed Societies were founded. Modern breeds are derived from two main types: Asiatic (e.g. Brahma, Cochin, Pekin, Malay) and Mediterranean (e.g. Ancona, Andalusian, Leghorn, Minorca). Other breeds were developed by crossing and selecting, their names often indicating their geographical origin. The Scots Grey, for example, originated in Scotland over 200 years ago, while the Sussex was listed in an English poultry show in 1845 and its make-up includes contributions from the Brahma, Cochin and Dorking. Similarly in America, Plymouth Rocks, Wyandottes, Rhode Island Reds and other breeds were being developed towards the end of the nineteenth century. In addition to the large fowls there are a number of bantams, which are miniature versions of the large breeds, often resembling them closely in conformation and plumage, but with a much smaller body size. There are 58 large fowl and 11 bantam breeds listed in the *British Poultry Standards Handbook* (May and Hawksworth, 1982), and 51 large fowl breeds and 62 breeds of bantams listed in the *American Standard of Perfection* (American Poultry Association, 1989). Additional breeds are recognized in other countries, together with numerous commercial hybrids, which are strains and lines rather than breeds.

The hybrids are of two main kinds: egg-laying and meat. The egg-laying types have been selected for egg number and size and can be further subdivided into two classes. Light hybrids, primarily from White Leghorn, with mature female body weights around 1.6 kg and white egg shells, are favoured in continental Europe and the USA, while medium hybrids, derived primarily from Rhode Island Reds, with body weights of about 2.2 kg, lay brown eggs and are especially popular in the UK (Figure 1.2). The meat-type hybrids originated from heavy breeds such as Cornish and White Plymouth Rock, and have been selected for growth rate and meat yield. Recently dwarf strains of meat-type origin have been selected – the dwarf gene is expressed in the breeding female, which has a lower food intake, is relatively small and is easier to manage, but not in the offspring, which grow much faster. Both egg-laying and meat types have been selected for 'efficiency' – maximum output for minimum food intake.

1.4   Domestication of other galliforms

The guinea fowl (*Numida meleagris*) has probably been domesticated for about 5000 years. The wild species is distributed across almost the whole of Africa

Figure 1.2.   Four important breeds of fowl from which most of our modern hybrids have been derived are shown. The first three are layers, the last is a meat breed from which broiler strains were selected. a. White Leghorn. b. Rhode Island Red. c. Light Sussex. d. Barred Plymouth Rock (May and Hawksworth, 1982). Female and male are shown for each breed.

a.

b.

c.

d.

south of the Sahara and is found over a wide range of terrain, though it is most common on savannah or in scrubland. Guinea fowl in ancient Egypt were maintained in garden aviaries by wealthy noblemen as an attractive feature. The first historical reference to them is found in an Egyptian mural dating from about 2400 BC and excellent representations appear on a Greek urn of the sixth century BC. They were well known in Rome by 30 BC, when both eggs and meat were regarded as delicacies (Belshaw, 1985).

Two species of quail have been domesticated: the Japanese quail (*Coturnix coturnix*) which has been divergently selected for egg production on the one hand and meat yield on the other, and the bobwhite quail (*Colinus virginianus*) in America.

The wild turkey (*Meleagris gallopavo*) is indigenous to the New World and was probably domesticated in Mexico in prehistoric times. It was imported to Europe soon after the early exploratory journeys to North and Central America, reaching Britain about 1525 (Thomson, 1964). There are several varieties (e.g. Norfolk Black, Bronze, White) and a number of commercial hybrids selected for increased meat yield, particularly from the breast muscles.

1.5   Domestication of ducks and geese

All domestic ducks, except the Muscovy, are descended from the wild mallard (*Anas platyrhynchos*), are given the same Latin name and were first domesticated in Southeast Asia or China at a very early date. Eighteen breeds are recognized in the UK (May and Hawksworth, 1982) and fourteen in America (American Poultry Association, 1989), varying in size from the Appleyard Bantam at 700 g to the Aylesbury at 4.6 kg, with some commercial hybrids weighing up to 8 kg. Most ducks are kept for meat production but some, such as the Khaki Campbell and Indian Runner, are extremely prolific egg layers.

The Muscovy (*Cairina moschata*) originated in South America, the domesticated form being similar to, but larger than, the wild species. Being of tropical rather than Palaearctic origin, it has a much lower body fat composition.

Domestic geese are descended from the Greylag (*Anser anser*); again they were domesticated in China or Southeast Asia, probably earlier than the duck. They were well known in Europe by 700 BC, as they are mentioned by the Greek poet Homer. The Crested Goose was valued by the Romans for guarding duties and reputedly saved the Capitol from the Gauls in 390 BC by raising the alarm (Thomson, 1964). Eleven breeds are listed by May and Hawksworth (1982) and by the American Poultry Association (1989).

1.6   Domestication of the pigeon

Domestic pigeons are derived from the Rock Dove (*Columba livia*) and have been domesticated for at least 5000 years. Images of pigeons dating from 3100

BC have been found in Egypt (Thomson, 1964). They were originally kept for eating; only later were they selected for their homing ability.

1.7   Biological features important in domestication

In these following sections the key features of avian structure and biology are described briefly, with special attention being drawn to those features of birds which have proved advantageous to domestication as well as to others which pose problems.

Birds were first domesticated for their behaviour – in order to be used for cockfighting – and behaviour is important in many other aspects of domestication. Hale (1975) has pointed out that species, both mammals and birds, which have been successfully domesticated share a number of common features – they form relatively large flocks and have a hierarchical structure with males affiliated to female groups. They show promiscuous matings, males are dominant over females and sexual signals are behavioural, rather than by colour markings or morphological structures. These features allow the animals to be easily managed in large numbers, while the maintenance of hierarchies through social dominance reduces the danger of injury caused by constant fighting. Promiscuous sexual behaviour allows any male to be mated with any female. Alterations to markings or structures often occur during selection programmes, so their irrelevance to successful mating is very helpful.

Parent–young interactions are important – favourable characteristics include a critical period of bond development such as imprinting, the acceptance by females of other young soon after hatching and precocial development of the young. This allows animals to be readily tamed because they can bond to man rather than to their parents and to be reared by surrogate mothers if required. Precocial development minimizes the length of time over which the young require specialized care and development.

Animals which show favourable responses to man, such as a short flight distance and little disturbance because of human activities are also well suited to domestication. Other behavioural characteristics which are helpful include flexible dietary requirements, adaptability to a wide range of different environments and limited agility. The domesticated birds do show many of these features.

Much of this book is about behaviour, and the nervous system and special senses are critical as a basis for behaviour. Later selection was for meat and then eggs, so an understanding of the muscular and reproductive systems is important. Feeding, ingestion and nutrient absorption are crucial to the level of production and performance which can be achieved, so the digestive system needs to be considered. Respiration and physical condition are two areas which can pose problems under intensive conditions, so it is necessary to consider the respiratory system and the integument. If further information is required there are several excellent books dealing with avian structure and function and with the biology of the domestic fowl (for example, Bell and Freeman, 1971; King and McLelland, 1975, 1979–85; Freeman, 1983–84).

1.8   Central nervous system

The avian brain has preserved many of the anatomical features seen in reptilian brains but there is a massive increase in size and complexity, especially in the cerebral hemispheres, optic lobes and cerebellum (Figure 1.3a). Even those birds with the lowest degree of brain organization have brain masses 6 to 11 times greater than those found in reptiles of similar body size. Birds do not possess the extensive neocortex seen in mammals and the majority of the cerebral hemisphere is composed of expanded striatal areas. The absence of neocortex does not, however, disadvantage the bird, as many of the functions of neocortical cells are assumed by striated neurones. With the complexities of flight and the importance of vision to birds, both the cerebellum and the optic lobes are correspondingly well developed.

1.9   Vision

Vision is important to birds and this is reflected in the fact that the avian eye is particularly large in relation to both the head and the brain: in the domestic fowl the eyes together weigh about the same as the brain (Figure 1.3b). The laterally placed eyes of herbivore and omnivore species like poultry extend over a visual field of more than 300°, but they cover a much smaller binocular zone than predatory carnivorous species with frontally directed eyes (Figure 1.4).

The eye is protected not only by mobile upper and lower eyelids but also by a nictitating membrane originating in the medial canthus and moving laterally across the eyeball. It sweeps the lachrimal secretion across the cornea, removing any excess on its return movement. It is transparent in diurnal species and impairs vision so little that it is suggested that birds fly with it extended, to protect the cornea and prevent it drying out.

Birds have good colour vision, diurnal species such as the fowl possessing more cones than rods. There is a central area, where the receptors are very closely packed to enhance optical resolution; it is circular in grain-eating birds but oval or band-shaped in water birds, probably to improve perception and recognition of objects on horizontal surfaces.

Projecting from the rear surface of the retina is a comb-shaped object called the pecten. Its function is unknown but it is larger in diurnal species and may reduce glare in bright light or increase diffusion of nutrients into the aqueous humour.

Experiments studying the visual perception of birds have shown that they respond to visual stimuli very much like human beings. If they are trained to

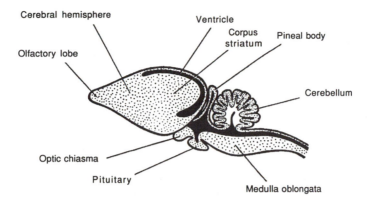

Cerebral hemisphere
Ventricle
Corpus striatum
Pineal body
Olfactory lobe
Cerebellum
Optic chiasma
Pituitary
Medulla oblongata

a.

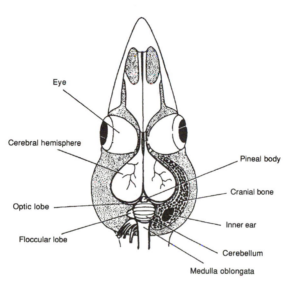

Eye
Cerebral hemisphere
Pineal body
Cranial bone
Optic lobe
Inner ear
Floccular lobe
Cerebellum
Medulla oblongata

b.

Figure 1.3. The avian brain. a. A longitudinal section of the brain to show the relatively large cerebral hemispheres, which co-ordinate the higher functions, and the cerebellum, concerned with the control of movement. b. The fowl's skull from above with the top removed to show how most of the space is occupied by the brain and the eyes (after Ede, 1964).

choose the larger of two objects and then offered a choice between two that are objectively identical in size but which differ in orientation or presentation so that one appears subjectively larger to human beings, birds too select the subjectively larger one.

This high standard of visual acuity and sensitivity is of little relevance to domestic birds kept in confined surroundings, such as hens in battery cages, but is much more important when they are housed under more extensive conditions.

There it helps them to identify sources of food and water, suitable nest sites and locations to scratch, dust bathe and roost. It is also crucial for individual recognition, for identifying sign stimuli and thus for the maintenance of dominance hierarchies and the social order.

1.10   Taste

The fowl has a well-developed sense of taste. The taste buds, averaging about 350 in number with a maximum of 500, are located on the dorsal surface of the tongue and in crypts at the openings of the salivary glands in the roof and the floor of the oropharynx. Behavioural studies (Halpern, 1962; Gentle, 1975) have

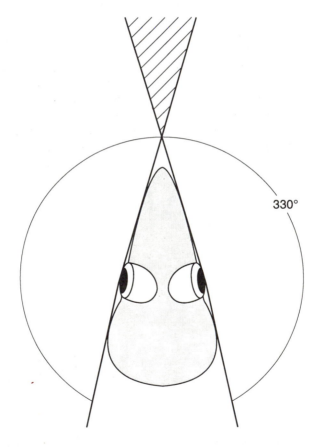

Figure 1.4.   Fowl's head from above showing how each eye has a very extensive field of view forwards, sideways and backwards, but that the area of binocular overlap (shown hatched) is relatively small.

shown that birds' responses to flavours differ in certain respects from those shown by mammals such as the rat. Acid and bitter flavours are rejected by the domestic fowl, as they are by the rat, but whereas rats prefer weak and moderate salt solutions to pure water, fowls do not. Both reject strongly saline solutions. Sweet flavours too, whether of natural or artificial origin, are generally not especially attractive to fowls, whereas they are selected strongly by rats. It may be that fowls use visual and tactile senses for food selection much more than do mammals, the primary function of taste being to reject items which may be noxious.

1.11 Hearing

The frequency range to which birds are sensitive is about 15 to 10 000 Hz (Bremond, 1963) and, like vision, this sense is generally very important for birds. Chicks of the jungle fowl, for example, when only 1 day old respond to long duration, high frequency sounds (such as the overhead predator call) by avoidance behaviour such as squatting down or running away (Kruijt, 1964). The domestic fowl has a large repertoire of about 20 separate and distinguishable calls, each given in a separate and definable context (Figure 1.5), which consist mainly of frequencies between 400 and 6000 Hz (Wood-Gush, 1971). Clearly, an acute and sensitive hearing ability is necessary to allow other birds to distinguish these calls accurately so they can make appropriate responses; auditory threshold tests on the fowl have shown that, while they respond at

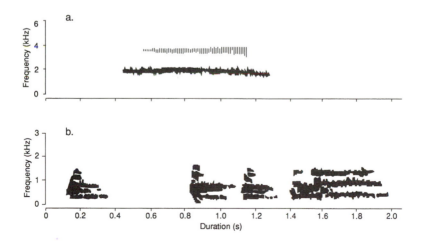

Figure 1.5.   Sound spectrographs of two of the domestic fowl's calls (after Wood-Gush, 1971). a. The overhead predator call is of high pitch (penetrates well), long duration (commands attention) and gradual onset and termination (difficult to localize). b. The ground predator call, in contrast, consists of a sequence of clucks.

frequencies from 250 to 8000 Hz, the range of most sensitive hearing lies between 3000 and 5000 Hz (Temple *et al.*, 1984).

1.12   Olfaction

Many species of birds have olfactory epithelium in the nasal cavity which is structurally similar to that found in mammals and it is possible to show positive neurophysiological reactions to various odours (Neuhaus, 1963). There is a clear correlation between olfactory bulb size and the extent to which particular species use olfactory cues in nature (Bang and Wenzel, 1985). Operant conditioning, field experiments and observations of behavioural responses to olfactory cues also provide convincing evidence of a well-developed sense of smell in birds (Bang and Wenzel, 1985). Kiwis have nostrils at the end of their upper mandibles and appear to use their sense of smell when probing in the ground for prey items. It is also known that, for migrating pigeons, odours carried on the wind can be an important navigational cue, while domestic fowls can be trained to respond selectively to particular scents, such as oil of citron (Jones and Gentle, 1985). For most bird species, olfaction is not as important as it is for mammals.

1.13   Cutaneous sensitivity

The skin of the bird is well supplied with sensory receptors, especially those areas of the body not covered by feathers, such as the beak. In the beak there are also concentrations of touch receptors grouped to form special beak tip organs, which allow the bird to make very fine tactile discriminations. Damage to the beak, of the kind imposed by beak trimming, will greatly impair birds' sensory abilities. There are three different types of touch receptors present: two which respond to a moving stimulus (Herbst and Grandry corpuscles) and one which responds to static pressure (slowly-adapting mechanoreceptors). Environmental temperature is monitored by cold receptors which respond to cooling of the skin, and warm receptors which respond to heat. Noxious (unpleasant or painful) stimulation is detected by another group of receptors, the nociceptors, of which there are at least three types present in the domestic fowl and which respond to either severe pressure or major changes in temperature.

1.14   Skeletal and muscular system

The lightweight bone structure of birds is an adaptation for flight. Storage of calcium in medullary bone is an adaptation to egg laying. The store is built up before birds come into lay and is then mobilized for egg-shell formation.

Development of massive pectoral muscles is another adaptation for flight. In largely flightless birds such as the domestic fowl, these muscles contain very little myoglobin and are therefore 'white'. Further selection has taken place during the development of meat-type strains in order to increase the mass of this desirable 'white' meat.

1.15   Reproductive system

The female avian reproductive system is unusual in that, although two gonads and oviducts begin developing in the embryo, those on the right side begin to regress quite early on while only the left ovary and oviduct continue to develop and become functional. The ovary grows especially rapidly with the onset of sexual maturity; in the domestic fowl it attains 60 g and occupies a mid-line position, overlapping the kidneys and lungs. It contains many thousands of oocytes, which develop sequentially into follicles. These grow very slowly up to about 2 mm diameter. A mechanism which is not fully understood then selects one follicle daily for rapid growth. It reaches the full size of 40 mm in about 8 days when it is ready for ovulation and, together with the other six or seven large follicles of various sizes, gives the ovary the appearance of a bunch of grapes (Figure 1.6). The primary oocytes of birds, the 'yolks' of eggs, are the largest cells in the animal kingdom, those of the domestic fowl each weighing about 20 g.

At ovulation the ovum is released into the abdominal cavity and is picked up by the funnel of the infundibulum (Figure 1.7). It passes down the oviduct into the magnum region, which has a deep glandular layer secreting albumen. This forms the thick coating of egg white around the ovum. It then passes through the isthmus, where the egg membranes are formed, and on into the shell gland (uterus). Here water is first transferred across the membranes, plumping the albumen, and calcium carbonate is then deposited to form the egg shell. The egg passes through the oviduct in about 25 h – about 20 of these are spent in the shell gland. Once oviposition begins, the sphincter between the shell gland and vagina relaxes, the shell gland contracts, the hen increases abdominal pressure and the egg passes through the vagina and cloaca and is laid.

After the membranes of the follicle have ruptured and released the ovum, its remnants form the post-ovulatory follicle. This has an important role – it secretes hormones (oestrogen and progesterone) which control the onset of pre-laying and nesting behaviour 24 h later, just prior to the laying of the egg (Wood-Gush and Gilbert, 1964, 1973).

Selection for egg number together with access to an *ad libitum* diet, has transformed *G. gallus* from the jungle fowl which, under natural conditions, lays a clutch of 10 to 20 eggs, through primitive varieties such as Indian village fowls typically laying 40 to 50 eggs in a year, to the modern laying hybrid. Its very highly developed oviduct, together with its liver where lipid for the yolk is

Figure 1.6.   Reproductive system of the female fowl: mature ovary showing five rapidly developing oocytes. The largest oocyte, on the right of the picture, will be next to be ovulated and the rest will be released at about 24 h intervals. A collapsed, postovulatory follicle is visible at lower left.

synthesized, produces over 300 eggs in 365 days. Initially ovulation occurs every 24 to 25 hours but as oviducal senescence takes place, the interval lengthens and sequences of eggs, which are separated by a non-laying day, become shorter.

1.16   Digestive system

The various species of domestic poultry, like all birds other than fossil species such as *Archaeopteryx*, lack teeth, having instead a horny beak with cutting edges. The tongue is heavily keratinized and is used for moving boluses of food within the oropharynx. Salivary glands are well developed – copious secretion from their numerous openings in the roof and floor of the oropharynx acts as a lubricant in swallowing the dry food typically consumed by gallinaceous birds.

At the lower end of the oesophagus is a dilated sac, the distensible crop, which is especially well developed in seed-eating birds such as the domestic fowl. Food stored in the crop softens and swells during storage, and is then moved on into the proventriculus lined with glandular cells, where true digestion begins (Figure 1.8).

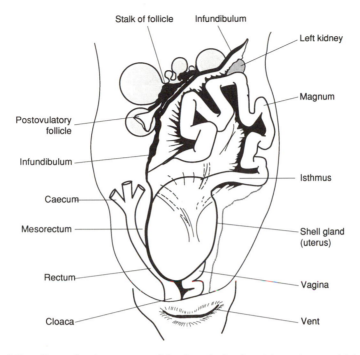

Figure 1.7. Reproductive system of the female fowl: oviduct, down which the egg will pass from the fimbria of the infundibulum to the shell gland and finally emerge from the cloaca (after King and McLelland, 1975).

The next stage takes place in the thick-walled muscular gizzard, which is lined with a hardened membrane. Under natural conditions birds eat small pieces of grit which localize here and the considerable pressures exerted by grinding movements remove hard seed coats and break down the food to small particles, which are digested by hydrochloric acid and pepsin in the gastric juice. Most modern poultry feeding programmes supply easily digested food, such as mash or pellets, and thus grit is not provided. Domestic fowls still have a tendency, however, to peck up and swallow whatever hard, indigestible fragments are available, which may explain why birds kept on litter sometimes pack their crops full with wood shavings.

Peristalsis then moves the food along the duodenum, where bile and pancreatic ducts open, into the jejunum and ileum, where the majority of absorption occurs. The large intestine consists of paired caeca and a short straight section, probably homologous to the mammalian rectum. Breakdown of food by symbiotic bacteria occurs in the caeca, while water is reabsorbed in the rectum and cloaca.

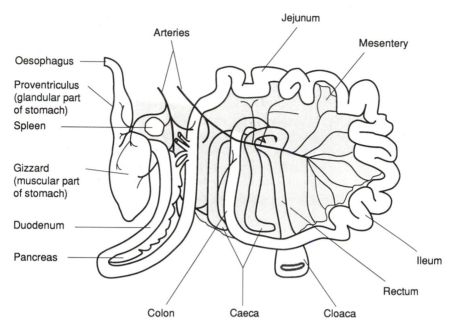

Figure 1.8.   Digestive system, showing the oesophagus, proventriculus, gizzard, duodenum, ileum, colon, caecum and rectum (after King and McLelland, 1975).

1.17   Respiratory system

Most birds inhale air through a nasal cavity containing complex, scrolled turbinate bones lined with mucous membrane. This acts as a system for recovering water from the moisture-laden exhaled air, with some of its water vapour condensing on the cool nasal membranes and being reabsorbed. This mechanism limits evaporative loss and is obviously an important adaptation to hot, arid conditions.

The larynx at the top of the trachea consists of four partly ossified cartilages supporting the glottis and prevents material such as food passing into the lower respiratory tract. In contrast to mammals it plays no part in producing the voice.

Because their forelimbs are specialized to form wings, birds have to use their bills for a wide range of functions, including investigative and aggressive pecking, preening and nest building. These functions require a long and mobile neck, which in turn necessitates a long trachea. To reduce resistance to air flow it is considerably broader than in mammals of the same size, so has a correspondingly greater dead space. This is compensated for by much slower rates of breathing and much greater tidal volumes than in comparable mammals.

At the lower end of the trachea, where it divides to form the primary bronchi, is the syrinx, where the walls of the air passages are formed by membranes stretched between circular cartilaginous rings. The function of the syrinx is to generate sounds, produced by vibration of the membranes, the tension of which can be altered by muscles which attach to the cartilages above and below them.

The lungs are in the dorsal part of the thoracic cavity, closely applied to the vertebral column and ribs, and are relatively small, with a volume only about a tenth of that of a similarly sized mammal (Figure 1.9). The primary bronchi divide into secondary bronchi and then into parabronchi, which in turn give rise to narrow passages called air capillaries. Gaseous exchange occurs mainly in the parabronchi and air capillaries. The latter have very small diameters, about 10 μm in swans down to 3 μm in small passerines. A network of blood capillaries runs along the walls of the air capillaries, separated from them by an extremely thin membrane, only 0.3 μm thick in the domestic fowl. There is a counter-current arrangement, with blood and air flowing past each other in different directions, which increases the effectiveness of oxygen transfer.

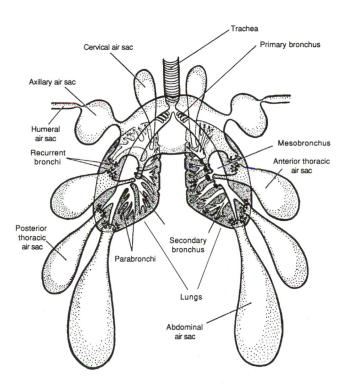

Figure 1.9.   Diagrammatic representation of the bird's respiratory system, showing the relationship between the lungs, the parabronchi and the air sacs, which occupy both the peritoneal cavity and some of the bones (after Ede, 1964).

The other major respect in which the respiratory system of birds differs from that of mammals is in their possession of air sacs. These arise directly from secondary bronchi and extend forward into the cervical region, throughout the thoracic region and well into the abdomen. The walls of the sacs have a minimal blood supply and their function is clearly not that of gaseous exchange but, rather, to control air flow. Diverticula from the air sacs extend into the cavities of a number of bones – in the fowl these aerated bones include the sternum, scapula, humerus, femur, pelvis and numerous vertebrae and ribs. Such lightweight bones may be especially liable to breakage if birds are handled roughly.

The avian lung is much more efficient than the mammalian lung, partly because of more rapid gaseous diffusion, and partly because the area of exchange surface is relatively far greater in birds than in mammals. The domestic fowl, for example, has 18 cm^2 of exchange surface per g body weight, in contrast to about 2 cm^2 in man. In addition, unlike the mammalian lung in which air flows back and forth, the avian lung has a unidirectional air flow. On inspiration air flows through the air capillaries into the anterior air sac and on expiration through the capillaries from the posterior sac (Schmidt-Nielsen, 1975). It is this high efficiency which enables birds to maintain energetic activities such as flying for long periods with such small volumes of lung tissue.

1.18   Integument

One of the clearest distinguishing features of birds is their feathers, which serve a number of functions: insulation, protection, waterproofing, cryptic coloration, sexual attraction and, not least, provision of the ability to fly. There are six types of feather: the most obvious, those which cover the outer surface of the body and include the flight feathers, are the contour feathers (Figure 1.10). They grow from feather follicles, initially as a richly vascularized dermal core. Around this an epidermal sheath develops to form the feather structure. This consists of a short basal tube, the calamus, which merges into the main shaft, or rachis. From this the barbs protrude at an angle and in turn the barbules arise from them, engaging by means of hooklets called barbicels with barbules from the adjacent barbs. Contour feathers cannot function effectively unless they are in first-class condition – regular preening ensures that barbules remain firmly interlocked, distributes sebaceous secretions from the preen gland and helps to rearrange disturbed plumage.

The other types of feather include down feathers, in which the barbules are not hooked, and which provide particularly effective insulation. Semiplumes have a similar structure but are much larger. At the base of each contour feather is a filoplume, which is richly innervated and probably provides proprioceptive input regarding the position of the feather. Bristles lack barbs, are found around the eyes and base of the beak, and have a tactile function. Powder feathers produce a white powder which helps to waterproof the contour feathers.

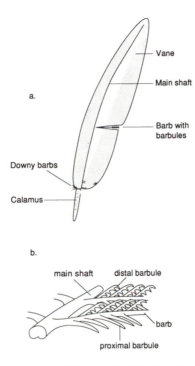

Vane

Main shaft

a.

Barb with
barbules

Downy barbs

Calamus

b.

main shaft   distal barbule

barb

proximal barbule

Figure 1.10.   Feather structure. a. Primary flight feather from the wing. b. Diagram showing how the barbs interlock by means of the hooked barbules (after King and McLelland, 1975).

Feathers deteriorate over time and are replaced by moulting. In domestic fowls there are three moults in the first 6 months, with an annual moult thereafter during the autumn if they are kept under natural lighting conditions. The moulted feather is extruded by growth of the epidermal layer between the dermal core and the calamus, until it emerges cleanly. If, however, it is pulled out at any other time, for example by another bird, the procedure is painful, the dermal papilla is damaged and the follicle fills with blood.

Beneath their feathers birds have a thinner and more delicate skin than mammals, which can be easily damaged if the protection provided by the feathers is lost. In many species, including the domestic fowl, the lower part of the hindlimbs is covered by scales rather than by feathers. On their heads fowls, as well as some other species, have comb and wattles, composed of richly vascularized tissue, which probably both serve an ornamental function, signalling a bird's status and condition, and help to dissipate heat. Birds have no sweat glands so their body temperature is regulated by evaporative cooling from the respiratory tract and by heat loss from the unfeathered areas. These two mechanisms are under behavioural control: panting increases evaporative cooling and wing raising exposes the poorly feathered sides of the body and under-wing area.

1.19   Application of biology to housing design

This brief survey of avian biology has emphasized that we are dealing here with complex, highly evolved animals. The different species of poultry have been studied in detail and a great deal is now known about their structure and their environmental, physiological and behavioural requirements. This presents us with the opportunity to put this knowledge to good use. It is important to take into account all the features of poultry when designing production systems – some can be modified and controlled, but others may need careful consideration if subsequent problems are to be avoided.

2

Poultry industry

2.1 Summary

- The development and form of the poultry industry reflect its objective of providing eggs and poultry meat for human consumption. However, they also have implications for the environment and for poultry welfare which may influence future changes.
- Many scientific and technological developments have been important in controlling factors which affect economic production. Scientific progress includes increased understanding and control of nutrition, disease and breeding. Technological advances include artificial incubation and brooding, manipulation of the environment (including photoperiod) and automation of husbandry procedures.
- The modern industry has distinct sectors, specializing in different species and types of bird. The egg production sector uses both single-stage and multistage systems. Birds for meat production are short-lived and usually kept in very large flocks. The breeding sector maintains selected lines, which are crossed to produce first and second generation hybrids for the other sectors.
- Specialization has resulted in an increase in size of units and a reduction in number of producers. Some producers are now multinational companies, growing their own food and marketing their own products.
- Future application of science and technology will include genetic engineering, environmental control (including electronic monitoring) and further automation. Some of these developments may be prejudicial to welfare, while others (such as mechanized collection of broilers) may be advantageous. The industry is likely to be increasingly affected by regulations and recommendations on welfare and on other environmental considerations.

2.2   Historical development

The modern poultry industry originated in the nineteenth century, though its roots, as described in Chapter 1, go back to Roman times. It developed progressively from numerous small flocks of dual purpose breeds scavenging around the barnyard, fed on scraps and home grown grains, and using natural lighting, incubation and brooding.

Along the road to large-scale intensive husbandry there have been a number of important scientific and technological developments, which have allowed all the factors important for economic production to be brought in turn under control. These included the artificial incubator and an effective brooding system, increased understanding of nutritional requirements and formulation of balanced diets, reduction and elimination of disease through improved hygiene, disinfection, vaccination and antibiotics, environmental and photoperiodic manipulation through housing design and the availability of electrical power and accurate control systems, and finally reduction of labour requirements by automation of husbandry procedures such as food and water supply, egg collection and manure removal. Of all livestock enterprises, poultry production is the most dependent on scientific knowledge and technological expertise.

2.3   Control of incubation and brooding

One of the earliest scientific developments, which enabled the industry to provide large numbers of day-old chicks whenever required, was the modern artificial incubator, which resulted from the invention of a simple and accurate thermostat in 1881 (Hewson, 1986), combined with gas-fired heating, effective control of humidity and a simple method of turning large numbers of eggs. Once the eggs had hatched, gas heated brooders kept the chicks warm. It took many years of development before artificial incubators were as successful, in terms of hatchability, as broody hens; indeed, bantams have been used to hatch small numbers of pheasant eggs almost up to the present day.

2.4   Control of nutrition

Developments in nutritional understanding started in 1912 with the discovery of vitamin A. Initially all poultry rations were home-mixed, but as the potential output of layers increased and the essential role of protein, vitamins and minerals became better understood, a complete, balanced diet could be formulated, which meant that birds no longer had to be given access to pasture and herbage in order to forage for themselves. By the 1930s a number of commercial firms had begun to supply compounded feedstuffs in mash or pelleted form, suitable for high-producing laying hens (Hewson, 1986). Diets were later tailored to meet particular needs: low-energy high-protein diets for growing birds, high-energy lower-protein diets for light hybrid layers and very high-energy high-protein diets for broilers.

2.5 Control of disease

As the scale of the industry increased and flock sizes expanded during the 1920s and 1930s, progress was delayed by a sharp increase in incidence of conditions like Marek's disease (Hewson, 1986), salmonellosis, avian tuberculosis and coccidiosis. In the middle 1930s Marek's disease alone could result in an annual mortality of 20% or more, while pullorum disease and fowl typhoid, because of possible egg transmission, posed a threat to human health. Poultry health schemes with systematic testing and slaughter of infected birds, together with selection for disease-resistant lines, were partially successful in limiting the problem. Later, suitable antibiotics were discovered in the 1950s and vaccines were developed on a commercial scale in the period from 1940 to 1970 against the remaining poultry diseases of importance such as Newcastle disease and infectious bronchitis (Biggs, 1990). With this development, and with the separation of birds from their droppings by putting them into battery cages, it became possible to house birds in close proximity in very large flocks. As late as 1967 the average flock size in the UK of cage-housed laying hens was only 2200 (Sainsbury, 1971). By 1988 the average size for all laying flocks was about 16 000 and 52% of birds were in flocks of 50 000 or more (MAFF, 1988). For complete control of disease an effective vaccination programme has to be coupled with an all-in all-out policy (section 3.8), where a site is completely cleared of birds so that houses and equipment can be thoroughly disinfected before a new flock is installed. This approach has resulted in a dramatic fall in mortality. In the UK in 1970 about 12% of hens would have died in a well-run laying flock between point-of-lay at 20 weeks and the end of the laying cycle at 72 weeks. The comparable figure in 1990 in well-managed units can be as low as 2–3%.

2.6 Control of photoperiod

Hens will lay well and persistently only when the day length exceeds 12 to 14 hours and laying flocks remained dependent on natural lighting in the UK and many parts of the USA until about 1945, when electricity became widely available on farms. This allowed day length during the winter to be extended by artificial lighting and rate of lay to be maintained at a high level throughout the year. It was subsequently found that rearing on a short photoperiod (such as 8 h light : 16 h dark) and then, at point of lay, increasing day length steadily and gradually over an extended period up to about 17 h light : 7 h dark gave the best results in maximizing egg size and number. In addition, to be able to provide eggs throughout the four seasons in the numbers required, it became necessary to bring pullets into lay in all months of the year. These two developments demanded a light-controlled environment, with the birds completely isolated from daylight (section 3.9). Recent advances include the use of ahemeral cycles in order to increase egg size at the expense of egg numbers.

2.7   Control of the bird's immediate environment

There are reasons other than photoperiodic control for intensive housing. In many parts of the world it can be sufficiently cold in winter to cause problems in outdoor systems with water freezing and hens with frostbitten combs, while losses of birds to various types of predators can occur and exposure to disease and parasites is often greater than in indoor systems.

Food costs make up about 70% of the cost of production, so keeping and feeding birds indoors greatly reduces food losses to wild birds and rodents. In addition, keeping birds in houses usually provides a warmer environment which markedly reduces food intake (section 3.6) resulting in improved efficiency of food utilization and thus reduced cost of production, especially in countries where cold weather is common.

2.8   Automation

Another improvement in economic production, coupled with the increase in scale, has been achieved by a progressive reduction in labour requirements through automating most of the basic procedures in all sections of the poultry industry. For example, provision of food and water is by automatic feeding and drinking systems. There has been a shift to battery cages for layers, from which manure is removed by scrapers or belts, or falls into a deep pit where it accumulates over the laying cycle. Many processing steps are done mechanically. Machines transfer eggs from incubators to hatchers, and eggs are collected by rolling on to a belt which conveys them to the end of the house and often on to cross conveyors to an egg packing machine.

Domestic fowls require substantial volumes of water: the daily consumption of a typical layer strain increases from 40 ml at day-old, to 120 ml at 6 weeks of age and to 180 ml or more once laying begins (Sainsbury, 1971). In battery cages water is generally supplied either via a nipple drinker or a drinking cup system – arranged so that each drinking point is between two cages (Figure 2.1). Every bird thus has access to two separate drinking points, a safeguard against dehydration should one of them malfunction. In floor systems drinkers are traditionally of the 'bell' type, so called because of their shape (Figure 2.2). One drinker can supply about 100 birds and as water is consumed the level is auto-

Figure 2.1. Drinking nipple and cup located at the boundary between two cages so that it can be reached by birds from either cage.

matically topped up, with the disadvantage that if spillage caused by movement or tilting occurs the water keeps flowing. This is a potential cause of wet litter which is one of the environmental factors with adverse consequences for welfare (Chapter 4). To overcome this, nipple and cup drinkers are increasingly being used in floor systems with broilers as well as layers.

Automated feeding systems are usually either chain feeders or, on deep litter, automatically filled hoppers supplying pan feeders (Figure 2.3). Such systems are often designed to present food as a shallow layer in the bottom of a fairly deep trough. This offers considerable advantages to producers because it reduces production costs by cutting food wastage but may cause feeding difficulties to birds which have been severely beak trimmed or have any other form of beak deformity. Automatically filled pan feeders are commonly used for broilers and turkeys.

Modified environment houses are ventilated by automatically controlled fans; under emergency conditions, such as power failure, there should be provision for fail-safe ventilation flaps which open under gravity when released.

2.9 Structure of the modern poultry industry

The facilities and expertise required for the different aspects of poultry production are now so specialized that the industry has become separated into distinct sectors. In all developed countries there has been a progressive reduction in the

Figure 2.2.   Bell-type drinkers commonly used in floor systems. These are readily cleaned and remain automatically topped up, but water is easily spilt should they be tipped.

number of producers and an increase in size of units. Some of these have become large multinational companies, often with a vertically integrated structure. This implies that their activities range from the growing of cereals and feedstuff formulation, through the breeding of specialized strains, the rearing of chicks, housing of laying flocks and production of broilers, to the marketing of eggs and finished meat products.

In some countries, for example Norway and Switzerland, there is a limit on the number of laying hens which can be kept on each farm. In most countries, however, the number of birds (layers or broilers) per farm has progressively increased. In the UK, for example, there were 141 holdings with over 50 000 laying hens in 1988 (MAFF, 1988) and these represented 52% of the total laying flock. In 1990 it was estimated that about 80% of the UK laying flock was in the hands of only about 300 production companies. The UK broiler flock totalled almost 75 million birds at any one time in 1988 (MAFF, 1988), which implies a total of about 450 million broilers grown in the UK in that year. In 1990 it was estimated that most of the broilers were in the hands of only about 12 integrated companies who supplied the poultry meat market from about 400 growing farms. In the USA, a similar progressive reduction in the number of producers and an increase in size of units has occurred. In 1990, the 20 largest broiler companies had 80% of the volume and about 60% of the laying flock was held by about 60 companies.

These developments have allowed large integrated production companies to benefit from economies of scale and the result has been very competitive consumer prices for eggs and poultry meat. There may also, correspondingly, have been a tendency at management level for birds to be regarded primarily as production units, rather than as sentient creatures with their own set of needs.

2.10   Egg production

The egg production sector consists of rearing and layer units. Chicks are hatched and housed either in rearing cages or on deep litter. The system can be single- or multi-stage (Figure 2.4). In a single-stage system chicks are brooded,

a.

b.

Figure 2.3.   Automated feeding systems. a. Chain feeder (and bell drinkers). b. Automatically filled food hopper.

either in cages or on the floor, and kept in the same brood-grow houses until they reach point-of-lay at 16 to 18 weeks. They are then transferred to layer houses. There are modern cage systems available, but not widely used, which are so flexible, with moveable floors, partitions, food troughs and nipple drinkers, that the birds can remain in the same location from 1-day-old to the end of lay. In a multi-stage system, once they no longer require the artificial heat of the brooder, the chicks are moved from the brooder houses to grower houses. At point-of-lay with both systems, the pullets are transferred to layer houses, either within an integrated unit or by being sold on from specialist rearing farms.

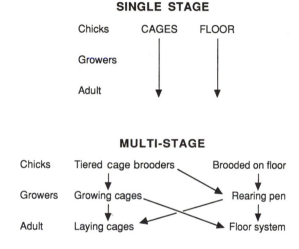

Figure 2.4.   Organization of egg production sector of the industry into rearing and layer units, which can be either single- or multi-stage.

Each system has advantages and disadvantages. In a single-stage system, because the birds remain in the same house to point-of-lay, it is believed to be less stressful and cause fewer checks to growth than when they are moved from one cage or pen to another (Sainsbury, 1971). However, it is much more difficult to provide accommodation which can meet the birds' requirements for suitable flooring, appropriate environmental temperatures and ready access to food and water from day-old to maturity than in a multi-stage system.

The majority of hens, both in Europe and the USA, are housed at point-of-lay in battery cages in groups of 4–6 birds. A small, but increasing, proportion is kept in non-cage systems – on the continent of Europe mainly in deep litter systems, in the UK in deep litter, perchery and aviary ('barn') systems or on free range.

Most flocks are slaughtered at 72 weeks because there is a gradual decrease in egg production and quality with time. As hens age, on average they lay fewer but larger eggs. By about 72 weeks of age egg production has decreased to 60 or 70% (Figure 2.5), although this is compensated for by an increase in egg size, so that total egg mass declines only slightly. However, there are also problems of egg quality with age. Egg shell thickness and strength decline, thus increasing the proportion of cracked and broken eggs. In addition, internal egg quality becomes poorer: the proportion of water in the albumen increases with age, resulting in unacceptably watery whites. In some cases, however, especially if the larger eggs laid by older birds are required, the flock is moulted, rested and brought back into production (section 3.9), which results in improved egg number, shell thickness and internal quality.

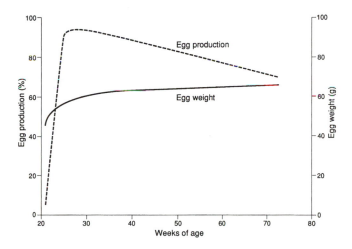

Figure 2.5.   Target egg production values for a medium-hybrid laying strain, showing hen–day production and mean egg weight from 20–72 weeks of age.

2.11   Meat production

The broiler industry began in the USA and reached the UK, in the form of both imported birds and management techniques, in the early 1950s.

Special strains and hybrids had been developed in the USA for meat production, selected for both rapid growth rate and an efficient food : gain ratio. They were derived primarily from two breeds: White Plymouth Rock and Cornish Game. Growth rate is so rapid that nowadays broiler chickens can grow from a 1-day-old weight of 45 g to a body weight of 2200 g by 42 – 45 days of age (Figure 2.6), when they are ready for slaughter and processing.

Flock size within houses varies from 5000 to 20 000. Many broiler sites consist of several such houses grouped together, for example 5 × 20 000 bird houses per site. The birds may be sexed at day-old and divided into separate

a.   b.

Figure 2.6.   Modern broiler strains are, with turkeys, proportionately the fastest-growing of all agricultural animals, from about 45 g at 1-day-old to 2200 g by 45 days of age, a 50-fold increase.

flocks of males and females. The cockerels require a higher-protein diet, and grow faster and more efficiently for a longer period than the pullets. The pullets thus reach the stage of diminishing returns earlier and are therefore processed and marketed at lower body weights than the males. The chicks are generally brooded in large groups under brooder temperatures of 35°C, with an initial ambient temperature of 29–30°C which is reduced by 3°C per week. A temporary barrier is sometimes used to prevent them wandering too far from the brooder. The behaviour of the chicks is an excellent guide to the correct combination of temperatures – they should neither huddle beneath the brooder nor arrange themselves at the periphery by the barrier or wall. The chickens are kept at very low light intensities of < 5 lux, partly to decrease the risk of outbreaks of feather pecking or cannibalism, but also partly to reduce locomotor activity, thus making a contribution towards improving the efficiency of food utilization. Stocking densities can be up to 20 birds per m^2 (Figure 2.7).

Almost all broiler birds are kept on littered floors in very large houses. The best litter is softwood shavings provided fresh for each flock. To reduce material and labour costs efforts have been made to develop suitable non-litter surfaces such as wire mesh or slats, but with limited success. A number of attempts have been made to house broilers in various types of multi-bird cage systems in order to simplify management, improve environmental control and reduce the cost of litter provision and disposal. These attempts have not been entirely suc-

Figure 2.7.   Broilers are stocked at high densities of up to 20 birds/m².

cessful, mainly because growing broilers in cages spend much of their time resting on their keels on the floor and from the localized pressure develop breast
blisters which result in carcass downgrading. Cushioned floors, where a rubber
or plastic overlay covers the wire mesh, reduce this problem. In some East
European countries and in the former Soviet Union caged broiler production is
common.

Turkeys are nowadays kept under very similar conditions to broilers, except
that turkey poults are sometimes reared in tier brooders, then moved to littered
pens at three weeks of age. Special attention must be paid to nutrition: poults
require a very high protein diet in the early stages and can be slow to begin
feeding. This problem can be reduced by positioning feeders under bright light
and also by including one or two domestic chickens in the flock. Their feeding
behaviour stimulates the poults to feed through social facilitation.

Ducks are also housed on deep litter, generally deep straw topped up daily
and usually with a slatted area next to the water troughs so that any excess
spillage can drain away. Cannibalism can be a problem and is often dealt with
by beak trimming. This is permitted in the UK on the same basis as for fowls,
that is, it should be carried out only when it is clear that more suffering would
be caused in the flock if it were not done (MAFF, 1987b).

2.12   Breeding sector

The breeding of poultry has become an industry in its own right. Genetic selection has made a major contribution to the success and development of modern poultry production and, because many of the economically important traits are governed in complex fashion by a combination of genes, the industry has greatly benefited from the emergence of the science of quantitative genetics. Breeding thus includes the development of new and improved strains, selected for a wide variety of characteristics. Egg-laying lines have been selected for traits such as egg number, egg size, shell strength, shell colour and low mortality. Meat lines have been selected for growth rate, meat yield, ratio of white to dark meat and rapid feathering. Recently there has been increased emphasis in both cases on selection for efficiency (high output:low food intake). These specially selected lines are then maintained by the breeder as grandparent stock. Because specialized sire and dam lines are unsuitable for general production purposes, they have to be crossed with each other to generate heterosis, or hybrid vigour. Commercial hybrids are thus often four-way crosses, being the product of parents which are themselves crosses between two inbred lines of grandparent stock (Figure 2.8). The eventual output of a breeder is therefore the supply of birds to the other two sectors.

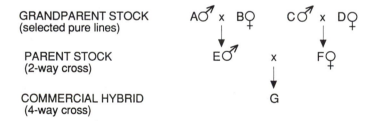

Figure 2.8.   Commercial hybrids are often four-way crosses derived from parent and grandparent stocks.

Considerable effort has been expended on attempts to intensify the breeding industry, but with less success than in the other sectors. Most of the broiler selection lines are still kept in small groups in floor pens, with natural mating. The parent stock are called broiler breeders and are kept to yield eggs which will be hatched to provide the growing broiler chickens. They are housed in larger flocks on deep litter, with one male per 9–10 females, and eggs are laid in tiered nest boxes from which they can be collected either mechanically or by hand. Cage housing and artificial insemination (AI) are used in some countries to a limited extent, but require considerable management skill.

Only in the case of turkeys has AI become the standard method and it was forced on the industry because selection for breast meat output resulted in such major changes in body conformation that natural mating became extremely difficult for the male.

2.13   Technical and scientific developments

Two areas which have begun to have major impacts on the way the industry may develop in the future are technical and scientific developments, considered here, and environmental and animal welfare considerations, introduced in the next section.

There have been recent advances in lighting technology and in the understanding of photoperiods. High efficiency fluorescent tubes are replacing tungsten sources, while the advent of intermittent and ahemeral light cycles (section 3.9) can reduce food costs or increase egg size and shell strength. In both cases the effects on bird performance have been carefully investigated but possible implications for behaviour and welfare have received much less attention.

Another field in which rapid progress is being made is that of molecular biology. It is technically more difficult to produce transgenic birds than transgenic mammals and so far genetic manipulation has had little impact at a practical level. One of the earliest applications is likely to be the transference of genes carrying resistance to disease.

The development of electronic monitoring and control systems, including the automatic weighing of broiler chickens (Figure 2.9), has introduced the possibility of an integrated management approach. This could, for example, involve the computerized control of ventilation systems and food distribution linked to changes in growth rate and egg output.

Recently developed equipment that enables the gentle, mechanized harvesting of broilers (Figure 2.10) offers advantages in terms of reduced stress and injury, with improved bird welfare and less carcass downgrading. However, the harvester is relatively bulky and its full potential can only be realized in modern houses with wide, unobstructed spans.

2.14   Environmental and welfare considerations

Pollution problems fall into two categories. If production units are sited too close to residential areas then there can be complaints about noise, dust and odours, especially from broiler houses in the last two weeks of the production cycle. These are problems which can sometimes be overcome by careful siting and good management. More long-term and serious are the problems posed by poultry waste and dead birds. The disposal of slurry from cage-housed layers by spreading it on agricultural land is causing concern because run-off into the ground or into streams and rivers may be one of the causes of increasing concentrations of nitrates in drinking water. Disposal of litter from broiler and

a

b

Figure 2.9.   Electronic perching platform for the automatic weighing of broiler chickens. Only one bird can stand on it at a time. a. General view of young broilers. b. Close up showing an older bird resting on the platform.

a

b

Figure 2.10. Mechanized broiler harvester. The broilers are drawn in by rotating rubber fingers and gently pushed on to a conveyor.
a. General view.
b. Pick-up head showing how rubber fingers interlock as they move inwards.

laying houses is also a problem. In the past, methods such as ensiling and feeding to cattle appeared to offer promise, but recent events involving disease transmission from one species to another mean that such approaches are unlikely to remain acceptable. Unless litter is entirely free from carcasses, the presence of botulinum toxin is also a very real danger. One possible method of disposal is indicated by recent research (Dagnall, 1989) which has shown that it is possible to compress the litter into fuel pellets. These have a high calorific value and can be burned to release heat and provide energy for electricity generation, leaving an ash which, being rich in potash and phosphate, is a valuable fertilizer. The disposal of dead birds also must be done in a manner that minimizes problems with odours, water pollution and spread of diseases.

Welfare regulations and recommendations are now beginning to direct the industry in terms of stocking densities, banning of mutilations, control of food restriction to induce moulting, setting atmospheric standards for dust and ammonia, and providing safeguards if automated control systems malfunction. There is also growing awareness of a requirement for first class management and stockmanship. This includes the need for attention to detail, the importance of a recording system (especially one which can provide an early warning of changes in variables such as daily water intake), the development of a feeling for stock and the ability to identify and interpret behavioural changes, and the ability to diagnose and quickly correct both mechanical and biological problems.

3

Environment

3.1 Summary

- Galliforms are an adaptable group, being predominantly ground-living birds, from complex tropical and temperate habitats. Jungle fowl live in dense rainforest and feral chickens make use of cover extensively, especially for roosting, emerging mainly to feed.
- Poultry systems have evolved from groups of a few birds in houses with nest sites and access to the outside, to large flocks in controlled environments. Modern systems offer many managemental advantages, including labour saving and hygiene, but generally provide simplified, even barren, conditions for the birds.
- The provision of food has also been simplified, most birds receiving a single, balanced diet, appropriate in quantity and composition to their state of production. Adult fowls are physiologically tolerant to a broad range of temperatures but temperature is closely controlled in most systems for economic reasons, to minimize food consumption.
- The need to maintain a minimum temperature of 21°C in cold climates without supplementary heating requires good insulation, large numbers of birds at high stocking densities and, on occasion, low rates of ventilation. This can compromise air quality by increasing dust, bacteria and ammonia levels, especially in floor systems.
- Manipulation of day length is used to control onset of lay, to prolong the laying period and to induce moulting. Intermittent cycles have commercial applications, especially for growing broilers and for layers; by increasing time spent resting they improve efficiency. Ahemeral cycles reduce egg numbers but increase egg size and shell strength. Light intensities in most systems are very low compared to natural lighting.

● Environmental conditions for poultry operatives must also be considered; air quality, ergonomics and the nature of the system are all important.

3.2   Natural and artificial environments

The Order Galliformes, which includes pheasants, quail, turkeys and fowl, is predominantly a ground-living group. However, even within a restricted group such as jungle fowl, which includes the ancestors and relatives of domestic hens, there is considerable variation in both habitat and range. This reflects an adaptability which is an important pre-condition for domestication (Hale, 1975).

In dense cover such as scrub, forest or jungle, wild birds are difficult to study and there have been few detailed reports of jungle fowl or other galliforms in the wild. However, there have been two important studies of feral chickens, with implications for the sort of habitat used by these birds. Both concerned populations on islands, one in the Great Barrier Reef of Australia (McBride *et al.*, 1969) and the other off the west coast of Scotland (Duncan *et al.*, 1978; Wood-Gush *et al.*, 1978). These studies emphasized the use of cover by the birds, especially for roosting in trees or bushes at night. They often fed in open areas, and short vegetation was particularly important for young chicks. Their habitat was, therefore, complex, and their adaptability has been exploited during domestication.

These wild ancestors and relatives of domestic poultry include both temperate and tropical species. In common with other animals, the temperate species are more seasonal in their breeding, because chicks born in the autumn or winter would not survive. Seasonality is a disadvantage under domestic conditions, but the sensitivity of poultry to certain seasonal changes has been exploited to increase production. This has been achieved by controlling light conditions and will be considered further below. Control of lighting is one aspect of the artificial environments that most poultry now experience and the adaptability of these birds can be seen in their response. Birds may breed and lay successfully in a wide variety of artificial environments, from cages to three-dimensional arrangements of perches.

3.3   Physical and social environments

In artificial conditions, as in the wild, different aspects of the environment interact with each other. The way in which birds relate to the structures around them is affected by social factors and vice versa. So although it is convenient to consider separately aspects of the physical environment (as in the following sections) and the social environment (Chapters 9 and 10), the two are not independent. In considering the physical environment of a domestic bird, social factors have both direct and indirect impacts. First, other birds are themselves an important part of the surroundings. It has been said that 'a sheep on its own is not really a sheep' and poultry are a similarly social species. The influence of other birds may be positive, as in reproductive behaviour, mutual preening and the formation of groups which feed or dust bathe together. It may be negative, as in aggression and feather pecking. It may also be physical, as when crowding leads to restriction of movement. Second, other birds modify the physical environment, affecting such factors as temperature, air quality, litter quality and disease risk. There are also interactions between direct and indirect impacts, as for example when birds roost close together for warmth. As one example of the interaction of physical and social aspects of the environment, we have already suggested that the best conditions for welfare are likely to be achieved in enriched environments housing small groups of birds at low stocking density (Figure 3.1). These different aspects are all discussed below where appropriate.

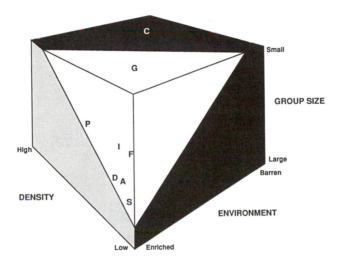

Figure 3.1.   Aspects of the environment interact in their effects on welfare. Here we suggested that welfare might be regarded as satisfactory in the clear part of the volume, increasing further with increasing distance above the interface. Placement of systems on the diagram was tentative. A, Aviary; C, cages; D, deep litter; F, free range; G, getaway cages; I, semi-intensive; P, perchery; S, strawyard (Appleby and Hughes, 1991).

3.4   Structure

Small scale, 'farmyard' poultry systems are in some respects similar to the wild, with a small group of birds, possibly with a male. These are provided with a house and roosts (equivalent to bushes and trees), a more reliable food source and nest boxes. There have been two main developments in larger systems:

larger group size (in floor systems) or increased control (in cages), or both. These developments have tended to involve housing, for protection and inspection of stock, for control of temperature and light and for reduction of labour. Decisions on housing depend on many factors, including climate. For example, closed houses enable modification of the environment round the birds which ordinarily results in, among other effects, improved food conversion efficiency, However, Curtis (1983), in the context of North American conditions, has suggested that in many cases totally closed houses are chosen primarily for labour saving convenience and worker comfort, rather than for animal shelter.

Large groups pose problems of damage to ground, hygiene and disposal of faeces, and hence the use of litter, slats or wire. Litter is provided for hens because keeping them on wet, droppings-coated concrete would be harmful for their feet. As with other species of farm animals, though, there has been a trend towards simplification of flock space, for both economy and control. In particular, facilities have been simplified. Perches have been seen as unnecessary for hens, and water (for swimming) as unnecessary for ducks. The importance of different facilities is covered under the behaviour appropriate to them in Chapters 8 to 12. The combination of large groups and simple, perhaps barren, environments means that in most systems except cages hens can move over a wide area. The effects of this, and the way in which it may be constrained by crowding, are considered in Chapter 12.

3.5   Food and water

Diets have also been simplified, in that poultry are normally fed a single, compounded, nutritionally balanced food. Its composition varies only with the developmental stage that the bird has reached: 'starter' diets are high in protein and vitamins, 'grower' diets lower in both and 'laying' diets high in calcium. Domestic poultry thus encounter food which is much less varied than that of wild birds and which takes much less time to eat. In addition, because it is usually provided *ad libitum* in the same location, it takes less time to find and effects of reduced foraging and feeding time are discussed in Chapter 8. One consequence is that heavy breeds of birds, for example broiler breeder hens, may over-eat and become excessively fat. Various forms of restriction have been tried to prevent this and to improve food conversion efficiency, including skip-a-day feeding, limiting the quantity offered daily, feeding in restricted periods and intermittent lighting.

Because a complete diet can be matched only to one particular level of production, it follows that birds producing at higher or lower levels will be underfed or overfed, respectively. This problem can be overcome by a self-selection dietary regime (Emmans, 1977). On small units with deep litter or straw, this principle has sometimes been put into action by part of the diet being left as whole grain, with the rest balanced as compounded mash. Grain scattered in litter encourages foraging by the birds and improves the condition of the litter.

Wet mash is more palatable than dry, but has to be mixed fresh. This is demanding of labour and is now rare. With dry food, domestic birds will drink more than wild ones, which eat a mixture of food items, many with a high water content. A reliable supply of fresh water will, of course, be particularly important in hot or dry conditions.

3.6 Temperature

As homeothermic animals, birds can maintain their body temperature over a wide range of ambient temperatures. In adult White Leghorn hens, this range is about −1 to 37°C (Esmay, 1978). Below this range, core temperature falls, while at higher temperatures it rises (Figure 3.2); perhaps surprisingly, the upper end of this range is less than body temperature (42°C in adult hens), but 37°C is nevertheless higher than would normally be encountered for more than a few hours. Within this range is a narrower range in which metabolic heat production is at or close to a minimum: the thermoneutral zone or comfort zone. Thermoregulation in this range is physical, including behavioural. Its limits are usually called the lower and upper critical temperatures, but these are not distinctly defined and estimates of them vary. They are also affected by variation in humidity (van Kampen, 1981) and probably by acclimation to particular temperature conditions. The thermoneutral zone for hens is somewhere between 20

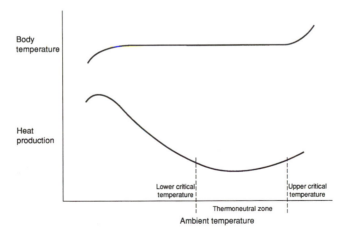

Figure 3.2.   Homeothermy. Below the thermoneutral zone (or comfort zone) the body burns energy to produce heat. Above this zone it must also use energy to stay cool and disperse the heat produced as well. At extreme ambient temperatures the body's mechanisms for warming or cooling itself are insufficient and its temperature falls or rises.

and 35°C, probably slightly above that for humans, which is 22 to 30°C (Esmay, 1978). Yet even outside this range there is no direct need for artificial temperature control, except in extreme conditions; with suitable diet, egg production is relatively constant across quite a wide range. This range has been estimated as from 15 to 27°C (Marsden *et al.*, 1987) or from 10 to 30°C (van Kampen, 1981). In temperate countries relatively simple housing can prevent extended periods of ambient temperatures outside at least the latter estimate of this range.

The main reason for close control of temperature is not productivity but food consumption. In cold conditions, a rise in heat production means higher food intake, so it is economic to keep temperature in the thermoneutral range. For laying hens, 21 to 24°C is usually recommended. Above this, food conversion efficiency may be further improved, but decline in egg weight may become a problem (Sainsbury, 1980) unless concentration of nutrients in the diet is increased. There have been many experiments comparing performance of hens at different temperatures. One example is illustrated in Figure 3.3.

For newly-hatched chicks, the temperature below which they cannot maintain homeothermy is about 26°C (Figure 3.4) and in normal, outdoor conditions they need to be regularly brooded by the mother (Wood-Gush *et al.*, 1978). Artificial brooding therefore requires higher temperatures, of about 30 to 32°C, than are necessary for adults even though the body temperature of chicks at 39°C is lower than that of adults. It is usual to decrease this temperature gradually, rather than subjecting birds to a sudden change. For growing birds, either layers or broilers, a lower target of around 21°C may be reached. As with egg production, the reason is not that growth rates would decline below this but that food conversion efficiency would decline.

In temperate countries, house temperatures adequate for adult birds can usually be maintained by insulation and wind proofing, without the need for supplementary heating. This will depend to a large extent on stocking density. The number of birds in a house influences temperature directly, through the heat that they produce, and indirectly, through the amount of ventilation which is necessary. At low densities, heat production is low and in cold weather in order to maintain an adequate ambient temperature the ventilation rate has to be low. This may give problems with air quality and with damp litter. Higher stocking densities allow more ventilation, but very high densities may necessitate ventilation so great that temperature control is difficult.

In countries with more extreme conditions, heating or cooling of houses, together with effective insulation, is frequently necessary. For best effect and most efficient use of energy, it is necessary to heat evenly the area occupied by the birds, which is most easily achieved by the use of hot air systems. Similarly, cooling systems that work on the air intake, such as foggers, misters and evaporative pads are most effective at cooling the whole house. Other direct measures which are utilized to prevent over-heating, including the use of open-sided houses, insulation of roofs to reduce downward radiation, the trickling of water

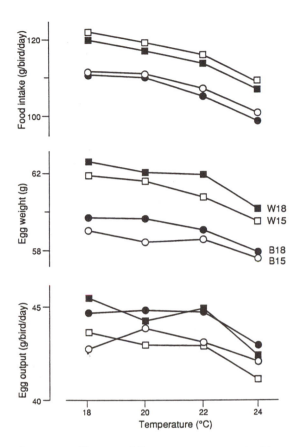

Figure 3.3.   Performance of hens at different temperatures. In this experiment, Warren (W) and Babcock (B) hens were fed on diets containing 18% or 15% protein (drawn from Table 3.5 of Emmans and Charles, 1977).

over roofs exposed to the sun and fans which increase air movement round the birds, will also be very important. In fact, orientation of buildings in relation to the sun and to other factors such as local winds has probably been under-utilized in tropical conditions (Smith, 1981).

3.7   Air quality

Hot air blowers may cause problems by drying the air. Relative humidity of up to about 60% is beneficial to growth in chicks (Sainsbury, 1980). Furthermore, respiratory infections are more likely in either dry air or very moist air, outside the range of about 40 to 80% humidity. High humidity can also give birds difficulty in keeping body temperature down, because in hot conditions body heat is dispersed mainly by panting and evaporative cooling.

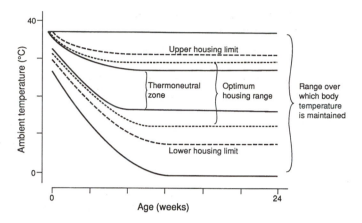

Figure 3.4.   Change in control of body temperature and in suitable housing temperatures with age, in White Leghorn chickens (after Esmay, 1978).

Respiratory infections are also increased by contaminants in the air, of which the most important are dust, bacteria and ammonia. Poultry breathing air with such contaminants develop lesions in the lungs (Maxwell *et al.*, 1989), which are associated with fluid accumulation and low blood oxygen (Odum *et al.*, 1987). These render birds more susceptible to infection (Anderson *et al.*, 1964; Oyetunde *et al.*, 1978). Problems with air quality are more common in floor systems than in cages, particularly where ventilation rates are low. For example, in one study of a deep litter house stocked at low density, average airborne dust was 30 mg/m^3 and average ammonia was 23 ppm; the birds were exposed to these levels over long periods (Appleby *et al.*, 1988b). The recommended maxima for short-term exposure in humans are 10 mg/m^3 for dust and 35 ppm for ammonia (Health and Safety Executive, 1980).

Air filtration is common in experimental houses but rare in commercial establishments. In an experimental study of its effects on broiler chickens, dust was halved and bacteria were almost eliminated (Carpenter *et al.*, 1986).

3.8   Hygiene and disease

Although there have been good reviews of the disease problems which may be encountered with commercial poultry, such as that by Sainsbury (1980), there have been few systematic studies of disease incidence. However, the risk of diseases spread by contact between birds, or by contact between birds and faeces, is generally regarded as more severe in non-cage systems. Certainly, one of the main advantages of cages is that birds are separated from their faeces. This is important in the control of diseases such as coccidiosis. It is supported by all-in all-out management, with houses thoroughly cleaned between flocks.

Similar management and cleaning are practised in other housing systems, but cleaning is, of course, not possible in systems incorporating pasture. Here the usual practice is to alternate or regularly change the area of ground in use, to prevent build-up of disease. In most free-range systems, birds are also fed inside the house to reduce the risk of contamination by pathogens derived from wild animals. Some contamination may still occur on range or from wild animals entering the house. However, if the flock is initially disease free, stockmanship is good, flock size is small and area per bird of pasture is large then the danger of land becoming 'fowl sick' (infected) may be small (Hughes and Dun, 1986).

Metabolic diseases which are not spread by infection have been reported at a higher incidence in cages than in other systems (Duncan, 1978), while skeletal problems such as osteoporosis are most serious in the confined conditions of cages (Rowland *et al.*, 1972).

3.9 Light

The primary biological rhythms in poultry, as in other animals, are seasonal and diurnal, both mediated by light. The main factor controlling seasonal changes in physiology and behaviour is day length. This control is at least partly mediated by the hormone melatonin, which is produced by the pineal gland primarily during darkness: production is suppressed by neural signals from the retina resulting from incident light. The concentration of melatonin circulating therefore declines in spring and this has an effect on sexual development. For day length to have its controlling effect it is necessary for the dark phase, or night, to be properly dark: light levels below 0.5 lux are recommended. To achieve this in housing when it is light outside, entry must be restricted to the light phase unless double doors are fitted and baffles are necessary around ventilators and other potential light sources.

Artificial control of day length for laying birds has two primary aims. First, it is an advantage to avoid them maturing too early, at too low a body weight. Birds maturing early lay small eggs, not just initially but throughout their life. This is avoided by use of a constant but short day length during rearing. Second, light control is used to bring birds into breeding condition and to keep them in this state for an extended period. This is achieved by a sharp increase in day length once the birds have reached their desired weight. In temperate countries, it is common to follow this by a slow increase (for example, of 20 min per week) until a day length of 16 or 17 h is reached. Any advantage of this approach over maintaining a constant, fairly long day is, however, probably minor. In hot climates, the latter system is usually the only one possible, because in houses which are not fully enclosed, the light period can be supplemented but not curtailed. Supplementing the natural daylight up to the longest day length of the year, or to a suitable longer period, avoids the danger of

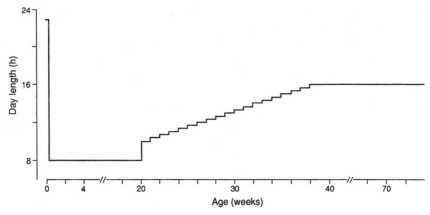

Figure 3.5.   A typical lighting programme for laying hens in an enclosed house.

decreasing production as day length declines. A typical lighting programme for laying hens is shown in Figure 3.5.

Despite the probable influence of melatonin in control of breeding condition, there is evidence for a direct effect of day length: of the separation of dawn and dusk as such. For example, 'dawn' can be indicated simply by a bright flash of light and such a flash coming 1 h before the start of an 8 h light phase may have the same stimulating effect as 9 h continuous light. This discovery has been utilized in the development of intermittent lighting regimes (Figure 3.6). In these, the light phase is interrupted by dark periods, which reduces electricity costs and may improve food conversion efficiency without decreasing egg production (Rowland, 1985). Some of the regimes investigated have been complicated, but it seems likely that most of the benefit to be gained might be achieved on a simple programme, for example with one or two dark periods during the light phase. It has also been argued that an increased food intake during what would

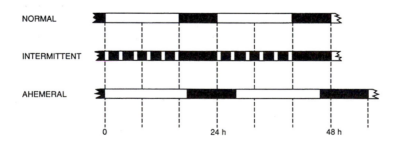

Figure 3.6.   Examples of different sorts of lighting regime. In the intermittent regime, a light phase of normal length is interrupted by dark periods. In an ahemeral regime, light and dark phases do not total 24 h.

normally be a prolonged dark period improves egg shell quality by allowing more constant absorption of calcium.

Intermittent lighting is also now used for meat birds. The main reasons are to reduce electricity costs and to improve efficiency of food utilization. Thus broiler chickens, which have usually been kept on a day length of 23 h to maximize food intake and growth rate, may grow as fast on an interrupted light schedule. Benefits in efficiency come from decreased activity during dark periods and perhaps from the reduction of 'boredom eating' in these birds which have little else to do, resulting in improved digestion.

One other way in which day length is used to influence laying is in the induction of moulting (Jensen, 1981), which is achieved by combining short days with food restriction (and sometimes water restriction, but see section 8.4). The objective is not the moulting itself, but termination of laying. After a break in laying, when day length is again increased, laying resumes at a greater rate than before. In the past it was common to keep flocks for more than one laying year, but nowadays induced moulting is mostly used to adjust the supply of eggs to the market and to rejuvenate egg shell quality, extending egg production by about 30 weeks.

In contrast to seasonal rhythms, diurnal rhythms do not necessarily involve day length. The main signal for controlling the pattern of behavioural and physiological variation over 24 h is usually dawn, although in continuous light or dark other factors can act as cues, such as feeding time, routine husbandry or temperature fluctuation. Whatever the main cue (or 'zeitgeber', German for timegiver), it controls the timing of such events as ovulation and laying. In hens, most laying occurs in a period of about 5 h starting shortly before dawn. Ovulation precedes laying by slightly more than 24 h (section 1.15), so it must therefore have occurred during the previous morning. The commonest pattern is for an egg to be laid early in the day, then others at intervals of 24 to 28 h, with fairly constant lag from day to day. When early afternoon is reached, a day is missed, bringing to an end the sequence; such a sequence is sometimes called a 'clutch', although there is little similarity between this and a natural clutch. After the missed day, laying begins early again. The proportion of days on which birds lay is determined by the length of sequences and hence the number of missed days. These relationships between the lighting pattern and laying are illustrated in Figure 3.7.

A hen with a lag of 2 h might lay eggs on successive days at 08.00, 10.00, 12.00 h, miss a day, then 08.00 h and so on, although such precision is unlikely. In this example, 18 h is 'wasted', from 14.00 on the missed day to 08.00 h on the next. It was this which gave rise to the idea of keeping hens on ahemeral lighting: use of light and dark phases which do not total 24 h. In theory, use of a 26 h cycle might use this hypothetical hen's time more efficiently. In fact, ahemeral cycles do not increase egg number; on the contrary, they decrease it (Shanawany, 1982). Cycles longer than 24 h do, however, increase egg weight and shell strength, mostly because the egg spends longer in the shell gland and

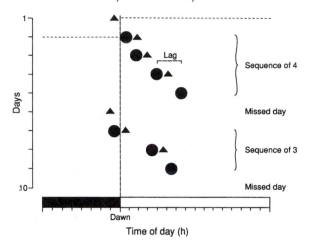

Figure 3.7. Example of ovulation and laying times for one hen. Ovulation (▲) is followed by laying (●) after about 24 or 25 h, as indicated for the first egg. Laying is then followed shortly by ovulation of the next ovum, except for the last egg of a sequence: when laying occurs late in the day it is not followed by ovulation so the next day is a missed one. In this example, seven eggs are laid in 10 days, so production is 70%.

achieves a thicker, heavier shell. These regimes can be used to increase the proportion of eggs in larger size categories, which may be worthwhile depending on price differentials. The main ahemeral system used is of 28 h (usually 18 h light to 10 h dark), because six such cycles fit into a normal week (Figure 3.6). Birds will entrain to this rhythm quite reliably – in other words their ovarian cycle will synchronize with it. One problem, though, is that day and night for the birds come at different times to those for the human workers. However, entrainment will also occur if the pattern uses bright and dim phases, rather than the usual light and dark, as long as the bright phase is 30 times the intensity of the dim one (Morris and Bhatti, 1978). Light in the dim phase can be adequate for work to be carried out, so husbandry can continue on a 24 h basis. It is not clear, though, to what extent behaviour also entrains to the ahemeral rhythm. There is some indication that a normal sleeping and resting period is absent on these regimes; in particular, oviposition generally occurs during the dark (or dim) period. Whether birds find this disturbing is not known.

Light intensity is important in itself. A certain minimum is necessary to stimulate ovarian function and so to maintain laying. This minimum is about 5 lux in hens (about 1% of average sunshine), and intensities of 10 lux or more are usually recommended. Turkeys need brighter light, of 20 lux or more. However, bright light also has various effects on behaviour which are adverse for either the owner or the bird. It increases activity, and probably for this reason tends to decrease growth (Cherry and Barwick, 1962), because activity uses energy. It also increases aggression and feather pecking. For these reasons the lowest practical intensities are used. Lighting for broilers may be as low as 0.2 lux, although supervision of stock is difficult at less than about 1 lux. In non-cage systems, it is important to avoid uneven light intensities, as these can result in uneven use of area.

It is sometimes suggested that it is important for birds to have access to natural light. There is no evidence to support this, but certainly some aspects of natural light variation help birds to adapt better to their environment. For example, gradual reduction of light at the end of the light period, rather than switching off lights abruptly, enables birds to feed in anticipation of the dark period, and to move to roost as appropriate. In partly open houses birds may react strongly to sunlight: if it shines only in restricted patches they often crowd into these areas, even to the extent of piling on top of each other (Gibson *et al.*, 1985).

3.10   Limits to adaptability

Poultry are adaptable to different environments, but there are limits to that adaptability. Some have been considered above, for example susceptibility to extremes of heat and cold. Others concerned with particular aspects of behaviour will be discussed elsewhere in this book and many of these also have environmental components. An example is that sometimes hens start eating their eggs. The causes are unclear, but environmental factors such as high light intensity and crowding contribute to the likelihood of egg eating occurring. Another example is feather pecking, which tends to be worse in barren conditions at high stocking density. These have been cited as examples of birds failing to adapt to their environment, which raises the question of what is meant by adaptability. Birds in natural conditions do not eat eggs or peck others excessively. However, birds which show such behaviour in artificial conditions are adapting, in that they are behaving in a way advantageous to themselves. Their behaviour is not advantageous to producers, but birds do not behave altruistically for the benefit of producers. The challenge is to design production systems in which birds acting for their own advantage will also be acting for the advantage of the producers.

3.11   Environment for operatives

Several of the aspects of the environment discussed in this chapter are important for human operatives working in poultry systems as well as for the poultry that they house. In particular, high concentrations of ammonia and dust in the air are unpleasant and potentially harmful, as indicated by the specification of maximum recommended exposure levels by the UK's Health and Safety Executive

(1980). There are two other major aspects of system design with impact on workers. First, ergonomics: the extent to which a system is comfortable and efficient to operate. Second, the general nature of the system: thus, many people work in the poultry industry because they like working with animals, and dislike intensive methods such as the use of battery cages. Conversely, some more extensive systems are unpleasant to work in; for example, some aviaries have platforms beside passages at head height, which allow birds to peck workers on the head.

Failure to take these factors into account in designing systems has frequently led to frustration, low morale and high turnover in staff. This is clearly a major matter for concern. In addition, it also has considerable impact on poultry welfare. If it is physically difficult to inspect birds in cages, birds are unlikely to be inspected properly. If a deep litter house has a high ammonia concentration, workers do not spend any longer than they have to inside. If workers are impatient and frustrated they are less likely to treat birds gently. Systems must be designed for humans as well as poultry.

4

Husbandry systems

4.1 Summary

- Poultry-keeping systems evolve over time: the earliest were farmyard operations, in which small flocks of birds had almost complete freedom of movement. There has been most development work on systems for hens, but many systems are used for various poultry.
- In free range systems the stocking rate must not exceed 1000 birds/hectare with a house providing deep litter or perchery conditions, where the birds feed, roost and nest. Consumers pay a premium on the grounds of enhanced welfare and for eggs which are perceived as having superior taste and nutritional properties. Free-range poultry meat can also command a premium.
- Strawyards, now usually covered, offer a compromise between outdoor and indoor conditions and house groups of 400–600 birds at ambient temperatures under natural ventilation and light.
- Deep litter systems usually have controlled temperature and ventilation, artificial lighting and a floor which, for layers and breeders, is part slats or wire mesh to reduce faecal contamination of the litter. Broilers are reared on a fully littered floor. Fully slatted or perforated floors, which slope to allow eggs to roll away, can result in behavioural problems, which are less if hens have access to nests and a sand bath, as in the Hans Kier system.
- Cages for laying hens in Europe with 4–6 birds at not less than 450 cm^2/bird, allow easy management at the expense of limiting movement and behavioural expression. In most countries they house over 90% of birds. Modified 'cages' usually have larger group sizes and more resources, such as perches, nest boxes, sand baths or litter.
- Aviaries and percheries are high-density, multi-level systems with large group

sizes. There are usually feeding troughs and nest boxes on each tier of slats or perches, roosts at a high level and litter on the floor.

4.2   Farmyard

Small scale operations in which hens, ducks or geese wander freely during the day are widespread on farms and rural properties. This is the only truly free range system, in that the birds are completely unrestricted in their movements, except that they are usually shut up at night for protection from predators. The house needs litter or bedding and may have roosts. With hens or ducks kept for eggs, nest boxes are also provided. Shutting birds in and letting them out takes some time; although they may forage widely, another daily task is providing supplementary food. These jobs may be labour intensive depending on the number of houses and birds, but the system falls outside normal economic analysis, since such labour is not costed and products are usually only for domestic use. Many of the principles and problems found in farmyard flocks recur in larger systems. For example, design and management of nest boxes for farm poultry, and remedies for birds failing to use them, have been considered over many decades (section 11.5). They continue to be important in all laying systems except battery cages. Even with small flocks, damage to ground is likely unless the house opens on to slats, concrete or gravel. Larger flocks cause more damage and with the development of commercial poultry keeping late in the nineteenth century there were two main adaptations: pasture-based, called free range, and house-based, called semi-intensive (Hewson, 1986).

4.3   Free range

In what used to be called free range, damage to ground and build up of disease were avoided by using small, moveable houses. These were still primarily for use at night, although additional light could be given inside to extend natural day length. Supplementary food was given in hoppers outside and the houses were shifted every few days or so to fresh pasture. Highly labour intensive, this approach is now almost extinct. It was adapted by the use of semi-fixed or fixed housing (Figure 4.1). Both of these provide a house big enough for birds to be fed inside, although they still obtain some nutrition from the pasture. The former involves a simple structure which may offer protection from wind and rain but little insulation, such as plastic sheeting over metal supports (Keeling *et al.*, 1988). This can be left in place for the laying year of a flock, then moved together with its nest boxes, feeders and drinkers. A move to another part of the same pasture may be sufficient, because one problem with birds on pasture is that they use the area near the house heavily but the rest very little. By moving the house, the area damaged during the year, and that under the house, is

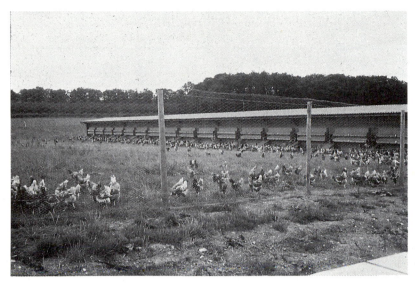

Figure 4.1.   Free range with fixed housing.

allowed to recover. Fixed housing does not allow this. Use of slats or other arrangements near the entrances to the house may reduce damage, but lower stocking density may still be necessary to prevent the development of unpleasant conditions. Good drainage of the ground away from the house is also important. With fixed housing, the system is effectively what used to be called semi-intensive. One traditional difference was that in free range a single area of pasture was permanently stocked, whereas in semi-intensive several areas of pasture were used in rotation. However, rotation of pasture (section 4.5) is now also used in free range. With permanent use of one area, stocking density must be low enough to prevent not only destruction of vegetation but also build up of disease (section 3.8). Some advisers recommend a maximum of 300 birds per hectare (Hann, 1980) and some up to 400 birds per hectare (Elson, 1985). However, European Community (EC) trading standards permit eggs to be sold as free range from flocks with stocking density up to 1000 birds per hectare. They specify, though, that the ground must be mainly covered with vegetation (Table 4.1; CEC, 1985). It is not clear whether it is possible to achieve this and to maintain healthy birds at such high densities over several laying years. Certainly this would demand good house design and high management ability, and in most cases lower stocking density must still be advisable for continuous occupancy of pasture. Even at lower density, the establishment of hard-wearing grasses is important and where possible the ground should be sown with an appropriate seed mixture.

For eggs to be described as free range, EC regulations also state that housing must have the size and facilities appropriate to deep litter or percheries (Table 4.1),

Table 4.1.   Criteria defined by EC trading standards regulations (CEC, 1985) for labelling of eggs.

Label	Criteria
a. Free range:	Hens have continuous daytime access to open-air runs
	The ground to which hens have access is mainly covered with vegetation
	The maximum stocking density is not greater than 1000 hens per hectare of ground available to the hens or one hen per 10 m^2
	The interior of the building must satisfy the conditions specified in (c) or (d)
b. Semi-intensive:	Hens have continuous daytime access to open-air runs
	The ground to which hens have access is mainly covered with vegetation
	The maximum stocking density is not greater than 4000 hens per hectare of ground available to the hens or one hen per 2.5 m^2
	The interior of the building must satisfy the conditions specified in (c) or (d)
c. Deep litter:	The maximum stocking density is not greater than seven hens per square metre of floor space available to the hens
	At least a third of this floor area is covered with a litter material such as straw, wood shavings, sand or turf
	A sufficiently large part of the floor area available to the hens is used for the collection of bird droppings
d. Perchery (barn):	The maximum stocking density is not greater than 25 hens per square metre of floor space in that part of the building available to the hens
	The interior of the building is fitted with perches of a length sufficient to ensure at least 15 cm of perch space for each hen

while in the USA there are no specific regulations. The cost of a free range system will equal the cost of one of those systems, plus land rental and fencing for the pasture, plus additional expenses. The latter will include higher food consumption (Hughes and Dun, 1986), and there will also be potential losses from factors such as dirty eggs, parasites and interference by predators. These increased costs are currently offset by increased income from the premium available on free range eggs; it remains to be seen how this premium holds up in future (section 5.7). In some EC countries this is the only system for which such a premium is available, since consumers apparently find the other categories in Table 4.1 confusing or unattractive. A proportion of consumers is prepared to pay more for free range eggs, but the reasons for this are not always clear. One reason is that hens on free range are presumed to have better welfare. At low stocking density and with good management this is probably a reasonable presumption, but these conditions are not always met. Another reason is the perception that free range eggs taste better, and are better nutritionally, than eggs from cages. Tests show that this is not true, except perhaps when birds are obtaining a significant amount of their diet from the pasture their eggs may taste different. A third reason became apparent during the recent period of concern in the UK about *Salmonella* contamination of eggs. Public opinion was that free range eggs should be healthier than those from 'unnatural' cages. If anything, the reverse is likely to be true. In some countries such as France, free range broilers and turkeys are common. This is for reasons of taste rather than welfare, since exercise will considerably affect muscle development. In this case, a premium for the free range product may have more justification. As with laying hens, use of the area of pasture available may be very uneven. For example, some flocks of turkeys on range in the USA have access to several hectares, but severe local crowding may occur.

4.4 Fold units

Fold units were an adaptation of the small, moveable houses used in early free range. Laying hens were enclosed in small pens, moved onto fresh ground every one or two days. An enclosed roost allowed a little more environmental control, but labour requirement was even more than for free range. This system is probably extinct.

4.5 Semi-intensive

Semi-intensive systems are primarily used for laying hens, and also offer housing and access to pasture. The house is fixed and, as the name implies, they traditionally had higher stocking density than free range. As mentioned above, this no longer excludes them from being called free range. One difference which used to be consistent in the two names was that in semi-intensive systems the higher density was supported by having two or more areas of pasture, opened in rotation. As mentioned above, this arrangement now also occurs in free range. One possible arrangement is to have a house surrounded by several such areas, with pop-holes for the birds next to each. Different pop-holes can then be opened as appropriate. Rotation is intended to prevent build up of disease and may also allow vegetation to recover. Alternatively, areas not in current use can be grazed by other animals, which helps to maintain a good turf. EC trading standards permit up to 4000 birds per hectare, for eggs to be called semi-intensive (Table 4.1) and this is the only difference from free range. It is unlikely, however, that this density can be sustained while retaining vegetation cover, even with rotation of different areas. As with free range, lower stocking density than this maximum is to be recommended. This system probably survives mainly for small domestic flocks, where it is more convenient to set aside a small amount of land for their exclusive use than to have them roaming at large.

4.6   Strawyard

Strawyards were introduced for laying hens as an intermediate between primarily outdoor and wholly indoor systems. Among other advantages, they avoided the need to shut up birds at night. They most often used an existing farm building, usually open-fronted, with an adjoining yard. The yard and part or all of the house had a thick bedding of straw as litter, with wire or slats in the rest of the house. Capital costs were low, but wet straw in open yards resulted in two main problems: build up of disease and high proportions of dirty eggs. Fully covered strawyards, also called covered yards, are now the rule (Figure 4.2). Similar houses are also used for turkeys, usually called pole barns. These may still be converted from existing buildings, or they may be purpose-built. They retain the same principle of using natural ventilation, with the upper part of the front of the house, or of several walls, being open. Solid lower parts to the walls, though, reduce draughts at bird height and screens can be used over the open-

Figure 4.2.   Part of a covered strawyard, with natural lighting and ventilation and a raised roosting platform.

ings in bad weather. Insulation is minimal, except sometimes over a roost area for increased protection at night, and the system is only suitable in temperate conditions where extreme temperatures are rare. Because the birds are within a house, though, supplementation of natural light is effective. Groups are generally of several hundred birds, for example of 400 to 600. Use of straw as bedding tends to limit this system to arable areas where straw is cheaply available, although other litter can also be used in similar houses. Straw may have some nutritional advantage over other litter. In one long-term study, birds in a strawyard ate slightly less than those in cages, with better conversion efficiency (Sainsbury, 1980). This must have been because they obtained some energy from grain remaining in the straw. In general, though, the increased activity in a yard and the low temperature which occurs for at least part of the time result in increased food consumption (Gibson *et al.*, 1988).

Under EC trading standards eggs from strawyards can be sold as deep litter eggs if stocking density is up to 7 hens per m^2 (Table 4.1). In practice, different authorities recommend less than 4 hens per m^2 (Sainsbury, 1980), or no more than 3 per m^2 (MAFF, 1987a). A study of densities up to 6 hens per m^2 found increased cannibalism and other problems above 4 per m^2 (Gibson *et al.*, 1988).

4.7   Deep litter

This system is not completely distinct from a strawyard, but it is usual for deep litter houses to be more fully enclosed. With powered ventilation, this allows more precise temperature control. In many cases, natural light is also excluded to allow the use of photoperiods shorter than day length. For this reason, deep litter is often used for the rearing of stock, even if they are to be housed in a different system later.

In any litter-based system, birds defecate on the litter and the consequences of this are important. Effects depend to a large extent on the behaviour of the birds, which is an integral feature of the functioning of the system. Thus the usual reaction of birds to loose litter is to peck and scratch in it. As a result, faeces do not simply accumulate but are dispersed. They may then dry out and be broken down by bacterial action. When this happens and the litter remains dry and friable, it is said to be 'working'. If litter becomes wet, however, because of water spillage, low temperature or high stocking density, it can also pack down and become solid. Either condition inhibits pecking and scratching, so a slight problem can rapidly become worse. Unpleasant conditions, including high ammonia, then develop and foot damage and disease are likely, so litter management is very important. The most common litter used is wood shavings.

Stocking density can be higher if part of the floor area is slats or wire mesh, so that the proportion of droppings accumulating in the litter is lower. This is achieved either by having a pit below the slats, or by having a raised, slatted platform (Figure 4.3), with droppings accumulating on the solid floor. Birds are

Figure 4.3.   An experimental deep litter house for laying hens. On the left is a slatted platform, with droppings accumulating below; on the right are raised nest boxes.

encouraged to roost on the slats at night, if necessary by moving them there nightly when they are first housed; this necessity can be avoided by rearing them with roosts (section 11.5). Drinkers placed over the slats reduce the risk of wet litter. With half the house area as slats, laying hens have been stocked at 11 per m^2 (Appleby *et al.*, 1988b) or higher, with no litter problems. Similar houses are also used for broiler breeders in North America and layer breeders in Europe. Broilers and European broiler breeders, however, are usually housed entirely on litter. Flock size varies with type of bird. In Europe, breeders are most often housed in groups of 4000 to 5000 and broilers in groups of 10 000 to 20 000. In North America, breeder houses usually hold 7000 to 8000 birds and broiler houses 12 000 to 25 000 birds.

Deep litter is also the most common system for housing breeding ducks. Straw is most often used as litter, because ducks drink a lot of water and produce wet droppings. These can be absorbed into open straw, but frequent renewal of litter is necessary. In contrast to the large flock sizes for hens, ducks are housed in breeding groups in small pens. Growing and breeder turkeys in the USA are also commonly kept in deep litter houses.

Sale of eggs as deep litter eggs in the EC limits stocking density of laying hens to 7 hens per m^2, with at least a third of the floor as litter (Table 4.1). At this density, though, the system is unlikely to be economic, while at higher densities there may be problems with restriction of movement or cannibalism (Appleby *et al.*, 1989).

4.8 Slatted or wire floor

Fully slatted or other perforated floors for laying or breeding hens allow a higher stocking density than deep litter, while reducing the risk of disease inherent in sustained contact between birds and their faeces. This system has often given rise to behavioural problems, though, including floor laying. In the Pennsylvania or Bressler system, therefore, the floor of the pens slopes down to one side so that floor eggs can be collected. Many, however, are inevitably broken. Other problems are cannibalism and hysteria, probably related to the combination of high density with a barren environment. These problems are apparently reduced, partly by the use of beak trimming, in new variants such as the Danish Hans Kier system (Figure 4.4). This has a sloping wire-mesh floor, with access to a sand area in the later part of the day when few birds are laying (Nørgaard-Nielsen, 1986). Flock size is about 80, compared with up to 200 in the original Pennsylvania system. Similar systems have also been tested in Switzerland (Matter and Oester, 1989). In the EC, however, eggs would have to be labelled as perchery eggs and in economic terms this system will compete poorly with multi-tier systems.

Figure 4.4. The Hans Kier system. The sloping wire-mesh floor has wooden perches above it.

Brooding and fattening of ducks is carried out on slatted or wire floors, because of the problems of maintaining deep litter in good condition.

4.9   Cages

Cages were originally introduced for single laying hens to allow recording of individual egg production and culling of poor layers. Later, several birds were placed in each cage, and group sizes of three to six are now most common (Figure 4.5). This resulted in considerable saving on capital costs and cages had other economic advantages, such as reduction of labour needed for maintenance and reduced food intake. The latter was partly caused by increased temperature resulting from higher stocking density in the house. These advantages have been further increased by the development of automatic methods for feeding, manure removal and egg-collecting. In addition, cages avoid some of the behavioural problems of high density floor systems, in two ways. First, the cage environment controls certain aspects of behaviour, such as egg-laying. Eggs are laid on the sloping floor of the cage and roll out for collection, so the failure to use nest boxes which sometimes occurs in floor systems becomes irrelevant. There are also adverse behavioural effects of such a restrictive environment, but these have only recently been considered seriously (Part II). Second, social problems associated with large group size, such as aggression and major outbreaks of cannibalism, are reduced. It is apparent that with regard to the welfare of laying hens, cages have both disadvantages and advantages. One major advantage is the separation of birds from their faeces and from litter, thus reducing disease and parasitic infections. Nevertheless, there do seem to be more welfare problems in conventional cages than in other systems (Appleby and Hughes, 1991).

Cages for larger groups of layers, called colony cages, were also tried at one time. They are now rare (but see next section). However, similar cages are also used for rearing domestic fowl and for housing small groups for mating if pedigrees are needed. In both cases, cage floors can be flat. Neither turkeys nor ducks are caged commercially, except for some breeding turkey hens, in which artificial insemination is routine.

Working conditions for operatives are often better with cages than with other systems: automation reduces the amount of physical labour, and dust and ammonia are usually less prevalent.

For any stock, food is supplied in a trough in front of the cage and water in an automatic system such as a nipple, cup or trough line through the cages (Figure 4.6). Recent improvements in cage design include the use of improved feeding systems to reduce food spillage, simplified cage fronts with horizontal bars which reduce feather abrasion during feeding (Elson, 1988a) and solid cage sides which also reduce feather damage (Tauson, 1989). In addition, faults in design causing birds to become trapped and suffer injury or death, which were previously a major problem (Tauson, 1985), are now much less prevalent

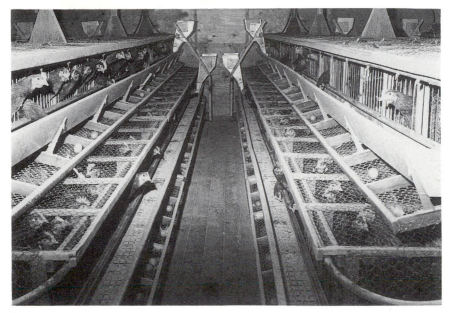

a.

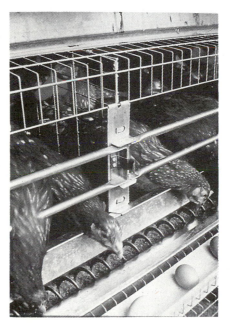

b.

Figure 4.5.   Types of cage. a. An old-fashioned cage house, with manual food distribution and egg collection, vertical-barred fronts and hexagonal wire-mesh floors. The cages are fully stepped over a deep pit; grids in the food troughs reduce wastage. b. Modern cages for laying hens. Simple doors are easy to operate and reduce feather wear; the chain which delivers food also reduces wastage.

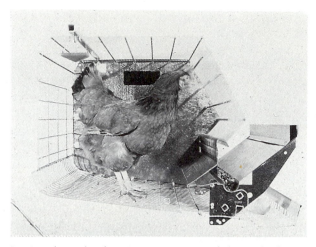

Figure 4.6.　Section through a laying cage. At upper left is a drinking nipple with a drip-cup; at lower right is the food trough, with a chain at the bottom which is pulled through to deliver food. Eggs roll under the food trough.

(Tauson, 1988). Cages are usually arranged in tiers. These are either vertically stacked, with faeces removed by a belt or a scraped shelf between the tiers, or arranged step-wise so that faeces fall into a pit sometimes below the house. The large number of cages in one house necessary to accommodate a flock came to be called a battery of cages and laying cages are still often called battery cages. It is common to have 20 000 birds per house in Europe and 60 000 per house in the USA. Space allowances for laying hens in cages vary in different countries from about 300 cm^2 per bird upwards, but the Commission of the EC has passed a Directive (86/113/EC) requiring its member countries to enforce a minimum of 450 cm^2 (Baker, 1988). The effect of any specific space allowance, however, will partly depend on the number of birds per cage, and hence the actual cage size. In the UK this is acknowledged in a further requirement that when there are only three hens per cage they shall have 550 cm^2 per bird, that two hens should have 750 cm^2 each and that a hen housed singly must have 1000 cm^2. There are further proposals in the EC of a minimum of 600 cm^2 per hen. Many manufacturers are now building cages of 2500 cm^2 for five birds, so that removal of one bird would meet the proposed increase in allowance (section 5.5). Effect of space allowance will also depend on other factors, including the size and activity of the hens concerned. For example, white egg layers are generally smaller than medium hybrids which lay brown eggs, so that recommended space allowances seem more generous. However, light hybrids are more active than medium, especially before laying (section 11.6), so the restrictions of the cage environment may be more important.

　　The EC Directive also specifies the minimum height for cages, of 40 cm over 65% of the minimum area and 35 cm for the remainder. It requires a minimum of 10 cm of feeding space per bird and a maximum floor slope of 8° for rectangular mesh. There are also regulations covering management, for example the requirement that all automatic and mechanical equipment should be inspected at least once daily. The emphasis of such legislation on caged birds reflects both the prevalence of this system (used for over 90% of layers) and its nature: in no other system are birds so intimately affected by every feature of the man-made environment.

4.10   Modified cages

Cage design continually evolves in attempts to improve economic performance, but there have also been attempts to modify design of laying cages to ameliorate welfare problems while not compromising economics more than necessary. The simplest of these is the shallow or reverse cage, in which the normal, narrow, deep dimensions of cages are reversed. This provides more feeding space, so that all birds can feed simultaneously. Birds tend to eat more, but also produce a greater egg mass (Hughes, 1983a). Capital costs are slightly increased, because more food trough is needed and because cage rows are narrower, which increases the proportion of the house which must be used for passages. Shallow cages, and those allowing brown birds at least 600 cm^2 each and white birds 500 cm^2, can also allow a perch across the width of the cage long enough for all birds to roost. Perches have negligible cost and encourage normal roosting behaviour (Tauson, 1984; Elson, 1985); in fact, they may reduce food intake (Braastad, 1990). Depending on design, they may also reduce foot problems and bone weakness (Hughes and Appleby, 1989; Duncan *et al.*, 1992). Their only drawback is that they tend to increase the number of eggs which are cracked or dirty. For this reason, some recent work on perches in cages has combined them with other facilities, particularly nest boxes (Robertson *et al.*, 1989; Appleby and Hughes, 1990). Modified cages, which provide such facilities while retaining the small group sizes and health advantages of conventional cages, are likely to provide more predictable improvements in welfare than more radical alternatives. However, the economics and practicality of providing such facilities for small groups have still to be assessed. Other modifications to conventional cages have involved larger groups. Colony cages in an unaugmented form are now rare, except for rearing and breeding, and an open-air equivalent called a verandah is now probably extinct. However, one recent study assessed a new colony 'cage' housing 20 birds with a litter floor, nest boxes and solid, transparent sides (King and Dun, 1984), but only on a small scale. Modern colony cages are also being developed in Switzerland, where conventional cages are banned from 1992, but 'cages' for 40 birds or more will be allowed (Matter and Oester, 1989).

The modified cage which has received most attention has been the get-away cage (Elson, 1981; Wegner, 1981). This incorporates perches and nest boxes and a greater freedom of movement vertically as well as horizontally, for groups of up to about 60 birds (Figure 4.7). Early versions had a flat floor and littered nests, but floor laying and dust-bathing in nests were problems, and sloping floors and rollaway nests have supervened (Wegner, 1990). One intention of the design was that subordinate birds should be able to use the perches for refuge from persecution, hence the name 'get-away'. Nevertheless, aggression and feather pecking have sometimes been severe, and there are hygiene problems because birds sometimes defecate on each other. Inspection and catching of birds are also more difficult than in conventional cages. Production results have been reasonable, but further improvements in design are necessary before this can be a viable system (Elson, 1985).

Figure 4.7   Demonstration get-away cage, with laying hens feeding and drinking from perches and a hen at lower left in a nest box. Normally each cage would house a group of up to 60 birds.

4.11   Aviary

The principle of allowing birds to use different levels, as in get-away cages, was then adapted to larger pens or whole houses. In the UK, aviaries were used first for breeding flocks of fowl, then adapted for layers. The aviary is basically a floor system, but with tiers of slats, wire or plastic mesh to increase the use of vertical space in the house. These tiers must either be strong enough for human use, or be arranged with passages for humans. Ladders between them are intended to encourage free movement of birds. Part of the floor is usually litter, although recent trials in Switzerland have investigated aviaries without litter (Amgarten and Mettler, 1989; Matter and Oester, 1989). Drinkers are placed over slats and feeders widely distributed. Nest boxes are placed as accessibly as possible, but floor laying is sometimes a problem (Hill, 1983). There are two other potential problems which have raised doubts as to whether the system is generally beneficial to poultry welfare. First, in common with other floor systems, litter can become wet and this has on occasion led to severe foot damage (Hill, 1986). Second, some birds can defecate on others. This may partly account for the intense feather-pecking which has occurred in some flocks, to the extent that birds were completely denuded (Hill, 1983). A recent development in Switzerland, Sweden, The Netherlands and the UK has been the use of manure belts under the upper tiers to prevent this problem.

Practical arrangements of tiers have allowed experimental stocking densities of up to 22 birds per m^2 of floor space (Elson, 1985). Within the EC, if the density is more than 7 per m^2 eggs must be sold as perchery eggs; otherwise they can be labelled as deep litter eggs (Table 4.1). Group size is usually of the order of 1000 birds.

4.12 Perchery

The perchery, a system for laying hens, uses the vertical space of houses by providing perches rather than platforms, arranged on a frame so that birds can jump up or down from perch to perch (Figure 4.8). Some experimental houses called aviaries have also used perches (Oester, 1986; Wegner, 1986), so the terms overlap. Similar overlap occurs in other European languages and some confusion is caused by the fact that EC trading standards (as in Table 4.1) use the term perchery in English, but the equivalent of the term aviary in other languages (for example, 'voliere' in French; perchery translates as 'perchoir'). Use of perches has allowed stocking densities of up to 18 birds per m^2 experimentally, with good results (Michie and Wilson, 1984; McLean *et al.*, 1986; Alvey, 1989). In these studies there was ample perch space and litter on part of the floor, which was used intensively at certain times. There were few behavioural problems (McLean *et al.*, 1986). However, the EC requirements for perchery eggs allow up to 25 birds per m^2 and do not include litter (Table 4.1). Without litter, birds will not use the floor fully and the minimal requirement of 150 mm of perch space per bird will not provide complete freedom of movement. Commercial firms applying these standards have encountered problems such as cannibalism (Harrison, 1989) and non-laying birds occupying nest boxes. Nevertheless, in some countries there is a market for eggs labelled as coming from this system, providing a premium for the producer. Early experimental percheries were pens with about 120 birds; subsequent versions contained flocks of 1250.

Figure 4.8.   Perchery for laying hens, with a framework of perches from which birds feed and drink. There is litter on the floor at lower right and several tiers of nest boxes at rear left.

4.13   Tiered wire floor

The tiered wire floor system developed in The Netherlands (Ehlhardt and Koolstra, 1984) is similar to the aviary in using tiers to increase vertical use of house space, but was designed with the requirement of matching the stocking density of three tiers of cages. The system resembles a cage house with the partitions removed: there are rows of narrow tiers with passages in between the rows and a manure belt under each tier (Figure 4.9). Nest boxes are provided, but the tiers are sloping in case eggs are laid on them. Perches are mounted over the top tier and food and water are supplied at all other levels except the floor, which is covered with litter. With stocking density up to 20 hens per m^2 of floor area, performance similar to that of caged birds has been recorded (COVP, 1988), although not consistently from flock to flock.

Tiered wire floor systems have already been adopted in Switzerland, where a ban on conventional battery cages is coming into effect earlier than in any other country (Oester and Frohlich, 1989). The Dutch design has been adapted in a number of ways, which are now being marketed in Switzerland and elsewhere under trade names such as 'Natura', 'Multifloor' and 'Voletage'. Nevertheless, problems such as floor laying, disease and cannibalism persist, and the management of stock is more difficult and demanding than in cages (Amgarten and Mettler, 1989).

The Elson Tiered Terrace (Figure 4.10) also has several wire-floored tiers and

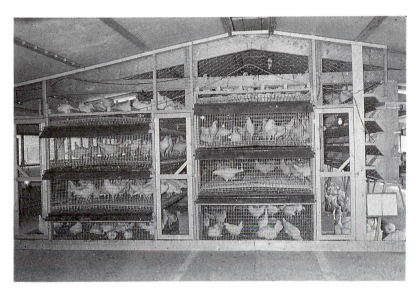

Figure 4.9. Tiered wire floor system, viewed from the end. The doors give access to passages between the tiers, for operatives. There is litter on the floor and there are nest boxes on the right.

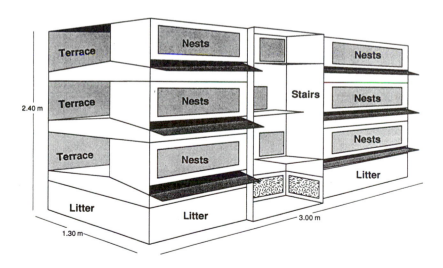

Figure 4.10. The Elson Tiered Terrace: a module for 150 birds.

loose material at ground level, but encloses these in modules for 150 birds (Elson, 1988b, 1989). Each tier has feeders, drinkers and rollaway nest boxes, and movement between them is by a stairwell. During the afternoon, this also gives access to the floor, which is covered with fine gravel. At other times of day, however, such access is prevented by one-way gates, which stop eggs being laid on the floor. Production in experimental modules has been reasonable, but feather pecking has been a problem. There has also been some mortality from ˙cannibalism (Elson, 1989). Further trials and development will investigate whether these problems can be eliminated and whether the system is viable on a commercial scale.

4.14   System design and poultry behaviour

Few developments in system design have taken the behaviour of poultry as a major premise. Rather, they have usually adapted previous systems in attempts to use currently available technology to improve economic performance. There have been two notable exceptions to this trend. Work on modified cages, first on the get-away cage and since then on other forms, was intended specifically to ameliorate the behavioural restriction imposed by conventional cages. Second, aviaries and percheries put into practice the suggestion of McBride (1970) that as wild fowl live in three dimensional conditions it should be possible to design housing in a similar way. Further developments of these two approaches seem most likely to provide appropriate matching between domestic poultry and their environment.

5

Economics

5.1 Summary

● The costs of production are lowest in standard laying cages, highest in free range (up to 70% more) and intermediate in other systems. Construction is simpler in non-cage systems, but low stocking densities result in higher costs per bird. More extensive systems also tend to have higher costs for labour and food, and greater losses from disease, mortality and unpredictable performance.

● Within systems, economic performance is dependent on standards of design and husbandry. Control of temperature and prevention of food wastage are important in reducing costs. Measures to prevent downgrading of eggs and injury to birds are important in safeguarding income.

● The market for eggs is inelastic: sales are not strongly influenced by price. However, profitability of both egg and broiler production is variable, with alternating periods of boom and bust. Short production cycles allow adjustment to short-term market trends, but tend to result in overcompensation.

● The availability of premiums for eggs or meat from poultry kept in extensive systems varies between countries. In some cases, premiums are sufficient to offset higher production costs and profitable specialist markets have developed.

● Achieving profitability in a fluctuating market involves long-term planning. Choosing between different systems is particularly difficult, involving decisions about the balance between capital and running expenditure and judgements about trends in public opinion (and hence availability of premiums). Possible changes in legislation on housing conditions must also be anticipated, although legislation is generally phased in slowly.

5.2 Comparison of systems for laying hens

For about 40 years, in most countries, laying cages have been the standard system of table egg production and over 95% of eggs have been produced by hens kept in battery cages. As described in the last chapter, several alternative systems came into use recently in which the eggs produced command a premium. Some of these are recent innovations, such as the perchery: others were in general use before laying cages became popular, such as deep litter and free range.

That costs of production vary considerably between systems was demonstrated by Elson (1985) who compared a wide range of husbandry systems, both commercial and experimental. Information was drawn from a range of sources and well-informed opinion was incorporated where necessary, to produce the results shown in Table 5.1. Taking the cost of production as 100 for birds housed in laying cages at 450 cm^2/bird (the minimum required in the EC under Council Directive 86/166/EEC, 1988) it will be seen that greater space allowances in cages, as well as production in different systems, increase costs. Allowing birds 750 cm^2/bird in cages increases production costs by about 15%, housing on deep litter by about 18% and on free range by about 50% (70% if the value of the land is included). Other systems are intermediate in cost. An extract of these results is also shown in Figure 5.1.

Table 5.1. Egg production costs in different systems for laying hens, relative to laying cages with 450 cm^2/bird. Space refers in cages to cage floor area, in houses to house floor area and in extensive systems to land area (from Elson, 1985).

System	Space	Relative cost (%)
Laying cage	450 cm^2/bird	100
Laying cage	560 cm^2/bird	105
Laying cage	750 cm^2/bird	115
Laying cage	450 cm^2/bird + perch	100
Laying cage	450 cm^2/bird + perch + nest	102
Shallow laying cage	450 cm^2/bird	102
Get-away cage, two tier		110
aviary	10–12 birds/m^2	115
Aviary	10–12 birds/m^2	115
Aviary and perchery		
and multi-tier housing	20 birds/m^2	105–108
Deep litter	7–10 birds/m^2	118
Strawyard	3 birds/m^2	130
Semi-intensive	1000 birds/ha	135 (140*)
Free range	400 birds/ha	150 (170*)

*Includes land rental.

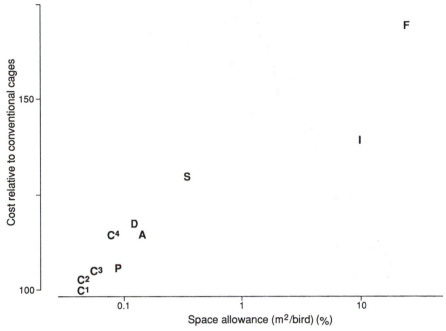

Figure 5.1. Cost of egg production in different systems (simplified from Table 5.1). Space allowance is plotted on a logarithmic scale. C^1 450 cm^2 cages; C^2 shallow cages; C^3 560 cm^2 cages; C^4 750 cm^2 cages; P, perchery; D, deep litter; A, aviary; S, strawyard; I, semi- intensive; F, free range (Appleby and Hughes, 1991).

In examining these costs of production, it is necessary to consider which factors influence them. These include housing, labour, food intake, hygiene, mortality and predictability of performance.

The simpler housing systems tend to be lower in construction costs. For example the strawyard system is uninsulated and has no fans and may be a cheap polebarn construction. However, stocking density is very low in such a system (3 to 5 birds/m^2) and therefore housing cost per bird is higher than in the more sophisticated, intensively stocked systems.

Labour is an important cost factor. Loose housing systems such as percheries and aviaries require more husbandry skill and a greater labour input. For example there are floor eggs to collect, birds to be trained to use nest boxes, treatment of pecked birds, nest boxes to be closed at night and so on. Dedicated staff, often prepared to work long hours in less favourable conditions, are required, which means increased labour costs.

In general terms the lower the stocking density in a system the higher the food intake requirement. This is because laying hens require a large proportion of their food energy intake for body maintenance which is directly affected by ambient temperature. It is well established (Emmans and Charles, 1977) that for every 1°C that the environmental temperature falls below the commonly recommended range of 21–24°C for laying hens food consumption is increased by about 1.5 g per bird per day. This is important in economic terms since food constitutes about 70% of total production costs. The more sophisticated, well insulated, controlled environment buildings containing intensive systems at higher stocking densities are therefore much more economic to run.

Hygiene is important for food safety and since in many alternative systems the birds are not separated from their droppings, care needs to be taken in them to avoid soiling of eggs and of other birds. This requires both design input and conscientious attention to detail to avoid problems of contamination. Regular and terminal cleaning and disinfection should be thorough to minimize the risk of disease outbreaks and egg shell contamination. Both these factors cause economic losses through reduced production and downgrading of soiled eggs. Where birds are in direct contact with their droppings, the risk of certain infections and internal parasitic infestations (e.g. coccidia and ascaridia) is considerably increased. This entails increased production costs due to extra staff vigilance to observe warning signs, labour to keep litter in good condition, medication and veterinary costs when necessary, and withdrawal of eggs from the market during and following treatment if drugs have to be used which result in harmful residues being left in eggs. There is also greater risk of certain external parasites, e.g. red mite, in some alternative systems. These also increase production costs through lowering the birds' performance and the need for treatment.

The death rate in birds housed in non-cage systems is generally, although not necessarily, higher than in caged layers. As mentioned above, the risk of certain infections and infestations is higher, and also feather pecking and cannibalism are often a problem introducing the dilemma of whether to beak-trim the stock. Such mortality often occurs in high-producing individual laying hens, which clearly entails economic losses.

Several other factors in alternative systems combine to make performance less predictable, thus increasing the risk of high production costs. One of these is the risk of floor eggs which are more costly to collect and are often dirty or cracked and therefore downgraded.

Similarly, feather pecking is very difficult to control and predict in several alternative systems, as are certain other conditions like ecto- and endoparasitic infections. Where low stocking densities are practised, the effects of weather conditions (especially humidity and wind) on the stock can be much greater and less controllable. All these factors add up to less predictable performance and in flocks where things go wrong, unpredictable increases in production costs.

As indicated in Table 5.1, the production of free range eggs, which generally attract a good premium, is considerably more costly than any other system, especially if the value of land is included. However, land can often be shared with other stock such as sheep or cattle. Also, land usually appreciates in value rather than depreciating like housing and equipment. Even if the land value is

left out, free range egg production still costs about 50% more than cage egg production. Some capital items, like high fencing to exclude predators, are expensive and certain running costs, especially the high food intake in winter, can be
considerable.

The cost of production in other systems falls between that for cages and free
range. It is about 8–15% higher in aviaries and percheries than cages, about
18% higher in deep litter and about 30% higher in strawyards. These findings
were confirmed by Haartsen and Elson (1989) who made a detailed analysis of
production costs in cages, aviaries and deep litter (Table 5.2). Tucker (1989)
compared costs of production in cages, perchery and free range systems. The
data are given in Table 5.3 and show costs in percheries of about 11% more,
and in free range of about 52% more than in cages (Figure 5.2). Clearly the
degree of management skill and attention to detail practised in any of these
alternative systems will affect production costs, and high standards of management and stock inspection are recommended.

Thus a wide variety of factors affect production costs. These factors are present to varying degrees in the range of poultry systems presented and should be
fully considered before deciding on a particular system. The effects of several
of these factors can be ameliorated by good management, but since labour costs
themselves are now high in many countries, husbandry skill and increased
labour requirements add to production costs.

Table 5.2. Production costs in three systems for laying hens: costs per hen
housed in Dutch guilders. At the time of writing 1 guilder = £0.30 = US$0.50 =
0.89 DM (from Haartsen and Elson, 1989).

	Cage	Aviary	Deep litter
Pullets (cage or floor reared)	7.50	8.00	8.00
Food	23.98	24.19	25.27
Animal health	0.03	0.10	0.10
Litter	–	0.10	0.35
Electricity	0.69	0.69	0.35
Water	0.12	0.12	0.12
Delivery costs	0.19	0.25	0.18
Bird depreciation	0.54	0.57	0.58
General costs (insurance, administration, etc.)	0.39	0.45	0.98
Housing costs	4.51	5.10	5.09
Labour costs	2.40	2.76	5.80
TOTAL	40.35	42.33	46.82
Slaughter revenue	2.07	2.26	2.38
Total costs per hen housed	38.28	40.07	44.44
Cost price per kg egg	2.06	2.16	2.39
Cost price per 100 eggs	12.76	13.35	14.77

Table 5.3. Performance and costs of production in three systems for laying hens. At the time of writing £1 = US$1.68 = 3.37 guilders = 3.01 DM. See also Figure 5.2 (after Tucker, 1989).

	Cage	Perchery	Free range
Performance			
Eggs per hen housed	276	265	252
Food intake (g per bird per day)	115	116	135
Mortality (%)	5	5	8
Old hen weight (kg)	2.18	2.18	2.27
No. of birds/poultry worker	20 000	10 000	2500
Costs of production (pence per dozen eggs)			
Food	25.6	27.8	32.8
Bird depreciation	7.9	8.4	8.6
Labour	1.5	3.2	13.3
Electricity	1.2	1.2	0.7
Medication	0.1	0.1	0. 2
Other costs	1.1	1.2	1.3
Total costs	37.4	41.8	56.9

Food: £140/tonne; pullets: £2.35 each; old hens: 24.2 p/kg

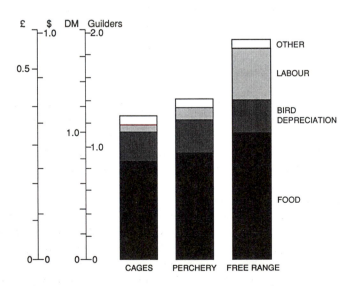

Figure 5.2.   Costs of production, per dozen eggs, in three systems for laying hens (from Table 5.3). Currency conversions are correct at the time of writing.

All the alternative systems to cages result in higher production costs, but as explained elsewhere in this chapter, premiums are available for eggs from certain systems in some countries. In countries such as Switzerland and Sweden cages are being phased out on welfare grounds, so consumers will be required to accept the higher costs, if poultry businesses are to remain profitable and therefore viable (but see Conclusions).

5.3 Comparison of poultry meat systems

Most broilers, turkeys and ducks are loose housed on litter and reared as rapidly as possible to obtain maximum growth rate and an optimum food conversion ratio. In some Eastern European countries and in the former Soviet Union, many broilers are grown in cages. Capital cost is higher for cages than for litter but these may be offset by savings achieved by rearing birds efficiently in small group sizes.

On the other hand a market has developed, especially in France and the UK, for slower grown, free range broilers and turkeys. This market is still in its infancy, but current indications suggest that free range production will expand rather like free range egg production has. Housing and production costs are considerably higher, because of a lower stocking density within the building and the fact that after 3 weeks of age the birds have access to the outside. Food consumption is also higher and food conversion deteriorates. However, the market is developing and it is possible that demand will exceed supply, thus ensuring a good premium and economic performance with profitable results.

5.4 Comparisons within systems

As we have seen, production costs are controlled to a considerable extent by the housing system employed. We have also noted that husbandry skill and attention to detail affect the performance of birds within each system. Thus, while no one would expect to produce free range eggs for the same price as eggs from laying hens in cages, free range production costs can be reduced by good design and conscientious husbandry.

In non-cage systems, many examples of good husbandry could be given, often as the application of common sense. For example, during cold windy weather the free range laying house will cool down rapidly if pop-holes are left open in exposed sites for long periods. The lower temperature in the house would cause the birds to eat more and increase the major factor in the cost of egg production, namely food. Good design of pop-holes so that they do not face the prevailing wind, and have baffles to avoid direct wind entry to the house, plus good management to ensure that they are closed promptly at dusk before the temperature drops too low, will minimize the problem.

Another example involves the production of naturally clean eggs. Since only first quality free range eggs can be expected to command a good premium, it is important to ensure that as many first quality eggs as possible are produced. Thus where litter is used it should be dry and clean and birds should enter the house from a dry area (under an overhang) round the pop-hole onto a perforated floor so that they clean up their feet during entry into the house. The nest boxes should be self-cleaning if of the rollaway type, or frequently inspected and cleaned if necessary where other types are used, and closed at night to prevent birds sleeping in them and defecating there. Floor eggs should also be minimized (section 11.5). This sort of attention to detail within any non-cage system will reduce downgrading losses due to dirty and contaminated eggs and ensure the production of a high proportion of valuable first quality ones.

Stocking density is another important economic factor. It affects food intake and therefore production cost. However, system design and management are also important, especially where high densities are used. Care is required in planning the layout of high density systems and operatives need to be especially vigilant to ensure that problems such as overcrowding in corners and floor laying, etc. do not occur, since they can rapidly lead to egg losses and mortality, which are both economic and welfare problems.

In laying cages also, care and attention to detail affect results and production costs. Two important factors in cages are the losses frequently experienced through food wastage and damaged eggs.

In the relatively barren environment of the laying cage, the food obviously provides much interest for the birds. If a sufficient depth of food is provided, birds will heap and flick it, which can be very wasteful. Food may be heaped in the back corner of the trough near to the birds and in piles along the length of the trough. Excessive food is used, both as direct waste, mainly over the back lip of the trough and through the cage floor onto the manure, where it is not very noticeable, and perhaps also as over-consumption when the birds take excessively large beakfuls because they are eating from a heap. Elson (1979) demonstrated that the delivery of food by conveyors at a low level to the base of the trough can lead to much more efficient feeding in cages than the previously used wasteful hopper trough method. These improved mechanized systems are now widely available, but good management is still necessary to achieve economy of food usage. The food level must be frequently checked, and adjusted if necessary, to ensure optimum food intake and food conversion efficiency.

In multi-bird laying cages, eggs may be collected manually or mechanically, using belt systems. In either case many eggs may be damaged when they roll down the cage floor into the collection area. Several eggs are generally allowed to arrive there before collection takes place and collisions occur as more eggs roll out onto those awaiting collection. The accepted norm for cracked eggs from cages is about 7 to 8%, the proportion increasing from only 1 to 2% from birds in early lay up to 10 to 15% from hens which have been in lay for a year. Well designed cage floors, together with good husbandry in the form of more frequent egg collections as the birds age, pay dividends. The proportion of damaged egg shells can be halved in this way and there is clearly a worthwhile financial saving since second quality eggs are worth only about one third that of first quality. Damage can be added in mechanical collection systems each time the conveyed eggs change level and direction. So there is an incentive to use one of the more efficient types of conveyor which handle eggs gently throughout the system.

A number of other cage design and management factors, although less important in direct financial terms, should not be forgotten. Often bird welfare and economics are closely linked as in the cases, for example, of foot condition and bone breakage. Tauson (1984) in Sweden has shown that the use of optimum sized floor mesh (or the use of a perch) and the provision of abrasive tapes mounted on the anti-egg-eating baffle plate, improve birds' foot condition and shorten their claws, thus reducing damage. This work has been repeated more recently in the UK (Elson, 1990) with similar results.

The problem of brittle bones in caged layers is dealt with in section 12.4. Gregory and Wilkins (1989) checked bone breakage on many end-of-lay hens as they were removed from cages and transported to the processing station, and found that gentle handling was vital to avoid excessive bone breakage problems. This is another example of a close link between bird welfare and economics, since hen processors spend much time and cost removing splintered bones from the poultry products produced from the end-of-lay hens. Good design (in the form of fully opening cage fronts) and care and attention to detail in handling the birds gently in and out of the cages will pay dividends.

5.5   Effects of legislation

Welfare legislation is gradually increasing in many countries and has a direct effect on production costs. Where eggs are produced in a particular system to meet market demand they can be expected to command a premium and therefore the extra costs of production should be more than covered. Where legislation requires producers to meet certain standards, however, production costs may well rise without any corresponding premium being available. In such cases, either consumers will have to pay more, or producers will have to absorb some of the extra cost, if that is possible, or stop producing eggs; alternatively eggs may be imported from surrounding countries which do not have such legislation, if that is permitted.

Where legislation effectively bans the use of laying cages, as in Switzerland from 1992 and in Sweden from 1998, the increase in production costs can be taken from Table 5.1. Sometimes, however, as in the case of Sweden, the basic cost of housing birds in cages is higher, since in that country from October 1989 birds in cages were required to have a minimum space of 600 cm^2/bird. Thus the increase in Sweden for producers going from cages to, say, an aviary system would be about 10% compared to 15% in countries starting from a lower base.

Meanwhile in many countries where cages are still allowed, legislation has been brought in to safeguard the welfare of birds. This legislation is covered fully in Chapter 6 and varies from country to country. Some measures like the fitting of an abrasive claw shortener currently required in Sweden have a minimal additional cost. Others, like the requirement to provide at least 600 cm^2/bird in Denmark or 550 cm^2/bird over 2 kg body weight in Germany, increase production costs by about 5%. Some legislation, like the Norwegian regulation that cage blocks should not be more than three tiers high, more directly affect housing cost.

The implementation of some legislation can limit the life of existing housing and equipment and therefore increase deadstock depreciation costs. For example, some cages installed in 1987 in EC countries did not fully meet the requirements of Council Directive 88/166/EEC (1988), which were implemented in 1988. This means that they will become obsolete in 1995, since all existing cages must comply by then. Such cages will be only 8 years old whereas their normal life might have been 15 years or more.

Another effect of changing legislation is that some manufacturers and producers anticipate it, so that they are fully prepared if and when it comes. For example, several groups are lobbying for a minimum cage space of 600 cm^2/bird in EC countries, and it is widely thought that legislation encompassing this will be enacted at some time in the future. Several cage manufacturers therefore offer a laying cage which has an area of either 2400 cm^2 or 2500 cm^2. Such a cage would meet the minimum requirement of 450 cm^2/bird in the current EC regulations by actually providing 480 cm^2/bird or 500 cm^2/bird when five birds per cage are housed. If one bird is removed, so that these cages each hold four birds, the area per bird is then 600 or 625 cm^2/bird respectively, thus meeting possible future regulations. Such a practice, whilst marginally increasing current housing costs (which are a fairly small part of total costs) provides some insurance against future legislation limiting the life of such cages. The risk is, of course, that the authorities may choose some other figure than 600 cm^2/bird or add other technical requirements that could not be met in these existing cages.

5.6   Egg market

The egg market tends to have cycles of over-production and low prices alternating with shortages and higher prices. The latter have been short lived in recent years because production can be stepped up fairly rapidly and because when prices are high eggs are imported from countries which have a surplus, if that is permitted.

The demand for eggs is inelastic. This means that reducing prices does not necessarily sell more eggs. Indeed there are some consumers who prefer to pay more with the feeling that by doing so they are getting the best. This aspect will be considered further in the following section.

The section on legislation showed how production costs can be affected when changes are made. Another factor is the effect such changes have on the volume of production. For example when a new rule was introduced in Denmark in 1988 that birds must have a minimum space in cages of 600 cm^2/bird, the national flock was reduced rapidly by about 20% because cages holding five birds could then only hold four birds. The effect of this was to change Denmark from being an egg exporting country to one importing eggs, mainly from Germany. Thus, changes in legislation which reduce a country's output have drastic effects on the market. In the short term they tend to increase prices and encourage imports. In the long term prices may be steady, but consumers have to pay more. Egg marketing in many countries is gradually changing. Traditionally, there were many types of outlet, including in the UK for example, milk rounds, grocers, butchers and fishmongers, farm shops, market stalls and supermarkets. Similarly, in the USA there were buyers, brokers (wholesalers and distributors), local grocery stores and supermarkets. More people buy milk in supermarkets now, so the milk rounds are under threat. If they go, then one outlet for eggs will also disappear. Increasingly, the majority of egg sales are through supermarkets. These supermarkets sell attractively packaged eggs and tend to deal with the larger egg producers who can supply them on a regional or nationwide basis.

Presentation and quality are important, especially in stores where eggs command good prices. Shells must be clean and intact, yolk colour deep enough to satisfy demand in the area and albumen firm enough to ensure that eggs stand up well when broken out. Packaging must be attractive and protective, and easily recognizable so that consumers will return for repeat orders, if they are satisfied. In most countries eggs are sold in several size categories, often divided at 5 g intervals, but in a few countries they are sold by weight.

The egg market and selling prices are organized slightly differently in various countries. Prices are, of course, controlled by supply and demand, but there are market leaders who tend to set the price which is then held if other traders follow. In The Netherlands there is a weekly auction which sets the price. In the UK, market leaders meet on Thursdays each week to set a guide price. Information flows very rapidly in modern communication networks so that, apart from slight regional variations, prices tend to be fairly uniform within countries over short periods.

Most countries now have a free egg market and in these, despite the efforts of the poultry industry to maintain them at what it regards as acceptable levels, egg prices are generally low in comparison with most other protein foods, so that eggs remain a worthwhile purchase. Certain countries operate quota systems which control egg output and in these, egg prices are generally higher, giving producers good margins. A recent example demonstrates clearly that a section of the public is willing to pay a premium for eggs that appeal to them as being the biggest and best available: large free range eggs are in great demand and command a large premium, whereas small free range eggs are sold as cage eggs with no premium.

5.7   Premiums for eggs from alternative systems

Probably due to a combination of increased public demand for eggs produced in systems which are perceived as enhancing birds' welfare, and a desire to buy the best eggs available even at premium prices, markets have emerged in some countries for eggs from non-cage or alternative systems. These systems are fully described in Chapter 4 and the effect of system on production cost was considered earlier in this chapter. The premium markets that have been developed to make viable the production of eggs at higher cost from some of these systems will be considered.

Specialist markets for eggs from alternative systems vary from country to country. Free range eggs have gained popularity in the UK and to some extent in France and Switzerland but not in The Netherlands, Germany or Scandinavian countries. Deep litter or 'scratching' eggs account for a few percent of egg production in Germany and The Netherlands, but 'barn' eggs from aviaries or percheries have not gained popularity in these countries until recently. In Switzerland the system of choice is the aviary and in Britain some eggs are produced in percheries and sold as barn eggs. In many countries like Canada and the USA where there is only a very limited welfare lobby, almost 100% of the eggs are produced in cages, although the production of free range eggs in the USA is now increasing to meet consumer demand.

Trading standards regulations which define the various systems available and describe their stocking density requirements apply in EC countries. These are detailed in Table 4.1. A description of how the market for eggs from alternative systems has developed in the UK, and especially in the south of England will illustrate the market possibilities and the potential for some other areas and countries.

In about 1980 the aviary system, which had been developed during the 1970s for breeding stock, was used for laying hens on a few farms. This was quickly followed by the perchery system which became more popular and 1–2% of the eggs were produced in this system in the early 1980s. The EC trading standards regulations then became effective and the eggs produced in the perchery system were sold as barn eggs (one of the EC definitions) at a premium of about 20% over cage eggs; this more than covered the extra costs of production. In the mid-1980s, the free range system was introduced in the south of England on a moderate scale and rapidly became popular at the expense of any further expansion of the percheries. Indeed, several of the farms that already had perchery houses installed free range systems as well, and within a few years 5% of the eggs produced in the UK were from flocks on free range and a further 1–2% from perchery birds with well over 90% still being produced by laying hens in cages.

During this period in the mid-1980s, some supermarkets were offering cage eggs, barn eggs and free range eggs. This was confusing to many consumers willing to pay a good premium and who, as we mentioned earlier, often want 'the biggest and the best'. On this basis barn eggs fell from favour, since when they were offered in stores at prices about 15–20% above cage eggs customers chose free range eggs at prices about 50% more than cage eggs. Demand for barn eggs therefore declined. Meanwhile the demand for free range eggs increased, especially for the larger sizes, and by the start of the 1990s had reached an estimated 10–12% of total production, with some producers expanding and many cage egg producers considering investment in the free range system. Egg packers and marketing companies have been predicting that demand for free range eggs may increase to 25 or 30% of the total market during the next few years, with cage egg production declining as a result. There has also, however, been a renewed increase in demand for barn eggs by some supermarkets that have ceased selling cage produced eggs.

It remains to be seen whether the above predictions are correct, but it is interesting to note that free range production was started in the south of England by someone who was looking for a specialist market opportunity. This rapidly became a success story, as described above, and free range egg production became, and has so far remained, a very profitable business; this shows the rewards that are available for enterprise in looking for specialist market opportunities and launching an appropriate new product. In this case it was a previously available product in which interest had been revived.

Cage produced eggs still constitute about 85% of total UK production, and the majority of consumers buy on price and are not concerned about the system of production. If cage eggs are to maintain or re-establish their share of the market, attention to quality, presentation, competitive pricing and a consumer education programme to answer concerns about bird welfare will no doubt be important. Efficient production of quality eggs at the lowest possible cost will continue to be vitally important.

5.8   Planning for the future

Forward planning is not easy in the poultry industry because of the supply fluctuations outlined earlier. The boom/bust cycles described seem to become longer in the bust period and have a shorter boom period as the years go by. However, planning ahead is important, as in any enterprise, and a number of factors need to be taken into account when assessing the future and making production and market decisions.

First, perhaps is the probable risk involved, both short and long term, and its predicted influence on profit potential. If short term risk is seen as a likely problem it would be wise to avoid investing until the risk period is over. Alternatively, a low cost investment might be considered during a period of likely risk possibly increasing investment in a more permanent system later. On the other hand a long term approach might be taken, in which case financial backers should be appraised of the risk and of the fact that extra help might be needed in the short term so that the business is in production and ready to cash in on higher returns in due course. Planned credit from a bank or financial institution is usually much cheaper than that arranged through a supplier of food or livestock when financial difficulties have already arisen.

Monthly statistics of chick placings are available in most countries and larger markets like the EC. Perusal of such information can provide a useful guide to likely increases or decreases in overall egg or broiler production a few months later. Techniques like induced moulting in countries where this is allowed can also be used to control egg output, or to take flocks out of production for a short period when over-production occurs and prices fall. However, care should be taken in applying this technique since there is an optimum age range when it is most effective. In general, a planned moult at an appropriate time is more successful than when the technique is used as a panic measure. The relative importance of capital versus running costs needs to be assessed. Sometimes a higher capital investment in a more sophisticated system will reduce running costs and therefore yield a greater return in the long term. However, high capital costs in a particular system could be a disaster in a country where such a system was under threat. For example, there is likely to be very little investment in laying cages in countries where a ban on cages is expected after a set period of years. Against that, a producer in such a country who foresaw the shortage of marketable cage eggs in the intervening years might continue to invest in cages knowing that the deadstock depreciation due to their short life would be high but expecting cage egg prices to increase during that period. On the other hand such a producer might invest in an alternative system and attempt to develop a premium market for the eggs produced, or even invest in broiler, turkey or other livestock production.

Very often new systems are installed in existing houses. A number of the perchery systems in the UK are housed in deep-pit laying houses, where the laying cages have become obsolete or needed replacing because of their age and condition. Where such constraints apply, decisions about the future of such a building have to be made. Assuming it is structurally sound and in good condition, it is a valuable asset and justifies further use. Whether to replace the cages with new and probably more efficient ones or to install an alternative egg production system or even a two-tier broiler production system, is an important decision. It is also important to ask how many tiers of cages the building will hold, whether cages should be installed upstairs and downstairs or just upstairs over a pit, or whether an aviary or perchery should be considered and if so, at what stocking density. Such questions bring into focus the value of housing and how it is best used.

Bearing in mind that the major cost of housing is in the roof structure, it is, of course, much cheaper to use the height of the building as efficiently as possible by installing as many tiers of cages or levels of perchery or floors in broiler systems as are viable. This is much more economic than building another house with half the tiers/levels and therefore half the number of birds in a given area of the building. The same principle applies when building new houses, so that the use of vertical space is more economic than increasing the area available. For this reason six tiers of cages with a catwalk halfway up (i.e. a double three-tier) are more cost effective and easier to inspect and operate than, say, standard four-tier blocks. For the same reason an aviary or perchery system, where the volume of the building is used more efficiently, is more economic than a deep litter system where only one level is used.

The decision on which system to invest in for the future will vary from country to country. As shown above, in some places where cages are likely to be permitted for the foreseeable future, restrictions on particular aspects such as stocking density may be added before too long. If this is likely to be within the life of current cage investments, then it would be prudent to take what advice is available in order to install cages now which are likely to be acceptable for their normally expected life span. In countries where a cage ban is threatened or proposed, any cage investment should only be made if the expected return can be shown to be good over a fairly short life. The alternative is to invest in another system and attempt to develop a premium market for the resulting product.

Another factor to be considered in planning for the future is sensitivity and risk. Companies do not just want the most profitable investment, but also the most reliable one. A strawyard system might become obsolete, for example, if straw became difficult to obtain. A particular system, especially if highly intensive, might rapidly become unpopular with the public in a certain area, and so on. Judgements must be made as to the best way forward in the light of recent experience and well informed future predictions.

Where such judgements are difficult to make there are sometimes decisions to be made over which choice of investment to make even within a system. For example, free range houses can be permanent controlled environment buildings surrounded by well designed and fenced pens, around which the birds rotate. Such a system would be high in capital cost but probably lower in running costs in the long term. Another approach to free range housing is to erect low cost plastic or straw bale structures which can be taken down and replaced on another site when the flock has completed its laying cycle. This approach would be lower in capital outlay and might well be justified if capital is limited, or if the market is currently profitable but risky, or expected to be short term.

5.9 Individual and flock profitability

Welfare pressures have increased in many countries in recent years and attention has therefore been focused on the individual well-being of the birds. It may be more important in the future to consider birds as individuals rather than flocks. This is an interesting approach because in individual terms economic performance and welfare are correlated; for example, reducing disease, mortality, bone breakage and feather loss are important to the hen in welfare terms, and to the farmer in economic terms. However, some practices which improve flock productivity tend to be at the expense of individual performance and welfare, so that increasing stocking density, for example, may depress individual output but give an overall higher return by reducing feed consumption and overhead costs.

Attention to individual needs in simpler conditions, and understanding and meeting better the behavioural requirements of individuals, may lead to better returns especially in an environment where consumers are prepared to pay more for products produced in ways which they perceive as being more humane.

6

Public opinion and legislation

6.1   Summary

- The way in which people regard animals is reflected in the laws which have been passed to protect those animals. Change in perceptions of animal welfare has been a gradual process, which has gathered pace since the early nineteenth century. Change in legislation has also been gradual, partly because legal definition of concepts such as cruelty and suffering has been difficult.
- Some countries have legislation which states generally that animals must not be ill-treated: this is often difficult to apply. Others prohibit specific actions, or prescribe particular conditions for animals, sometimes backed up by detailed codes of practice. Strictness of enforcement, however, also varies.
- Legislation in European countries is influenced by the Council of Europe, which produced a Convention on protection of animals, and by the EC, which imposes Directives on its members, and defines Trading Standards with specific requirements for husbandry systems.
- Public opinion in many countries is generally antipathetic to intensive husbandry. This is partly reflected by the fact that welfare societies are often more vocal in expressing their views than, for example, spokesmen for the agricultural industry. However, this stance is often inconsistent, since only a minority of the public is willing to pay more for poultry and eggs from extensive systems.
- Most people do not have a clearly thought out basis for their views on welfare, but adopt a compromise between use of animals and animal suffering. Legislation must, however, take into account ethical, sociological and scientific factors. In future, these may lead to legislation becoming prescriptive rather than proscriptive: defining how animals should be kept rather than how they should not be kept.

6.2 Animal welfare legislation in the UK

The way in which people regard the animals they keep for food, companionship, experimental purposes or sport is reflected in the laws which have been passed to ensure the protection of those animals. A historical approach helps to put into perspective what is happening today. Statutory legislation started as early as Anglo-Saxon times with a law of King Ine in 688–695. In more modern times the UK was a leader in this field: in 1822 Parliament passed a Bill forbidding the ill-treatment of horses and cattle at the instigation of Richard 'Humanity' Martin, Member of Parliament for Galway, against the ridicule of many of the other Members of Parliament. In 1835, Princess Victoria became patron of the newly founded Society for the Prevention of Cruelty to Animals, and a second law was enacted giving some additional protection to all domestic animals and outlawing bear-, badger- and bull-baiting, and dog- and cock-fighting. In 1876 the Cruelty to Animals Act was passed to give protection to animals in scientific experiments. All statute law must be applied using case law. The courts have been deciding cases concerning animals from time immemorial. Cases such as *Wilson* v. *Johnson* (1874) and *Everitt* v. *Davies* (1878) are still good law today. No statute by itself decides the law. To take an example away from the animal welfare context, such a basic statute as the Theft Act 1968 which in essence says 'Thou shalt not steal', is interpreted by lawyers in the light of many decided cases.

The most far-reaching legal milestone relevant to the welfare of farm animals in the UK was the Protection of Animals Act 1911. This Act established the criminal offence of cruelty to domestic animals, as evidenced by the infliction of unnecessary pain and suffering, and it remains the basis of UK animal welfare legislation today. From the animal's viewpoint it represented a major step forward, for the intention or otherwise to commit an act or omission which involved cruelty was immaterial; the only question was whether pain or suffering was inflicted and, if so, with good reason. Six classes of offence were recognized.

1. To beat, kick, ill-treat, over-ride, over-drive, over-load, torture, infuriate or terrify any animal.
2. To cause unnecessary suffering by doing or omitting to do any act.
3. To convey an animal in a manner as to cause it unnecessary suffering.
4. To perform any operation without due care or humanity.
5. The fighting or baiting of any animal.
6. The administration of any poisonous or injurious drug to any animal.

It is worth noting that the Act specifically included mental as well as physical suffering, by including the words 'infuriate' and 'terrify'.

Although at first sight the Act appears rather negative and specific, section 2, which makes it illegal to cause unnecessary suffering by doing, or omitting to do, any act, does give it a rather more wide-ranging impact. This section would, for instance, cover failure to provide for the needs of the animal, including food, water or suitable accommodation. The Act, which has been amended and extended on a number of subsequent occasions, was thus framed to control specific forms of cruelty and abuse in the context of early twentieth century farming and animal-keeping practices. It undoubtedly had a major effect on public attitudes, both by directly deterring abuse and also by changing people's perception of what was acceptable treatment of animals. The Act has, however, been criticized (Todd, 1989) for using terms like 'unnecessary suffering' or 'due care and humanity' on the grounds that they introduce such a major element of subjectivity as to make it difficult for a Court of Law to decide what is, and what is not, cruel. Those offering such criticisms should ask themselves, 'if a court cannot decide, then who can?' Court decisions are made by ordinary people: magistrates who, as a group, provide a vast pool of common sense, juries, or judges. Juries are capable of perverse verdicts but on the whole if presented with convincing evidence are not slow to convict, particularly where cruelty to animals is concerned. Judges, though sometimes maligned, have a vast knowledge and again are no friends to those who ill-treat animals. Persons who have brought unsuccessful prosecutions on animal welfare matters would do well to level their complaints not against the statute but against their evidence as presented to the courts. A most important point is that Magistrates' Courts are not Courts of Record as are the higher courts and it is therefore worthwhile to consider whether test cases on particular areas of the interpretation of statute law could not usefully be brought by animal welfare bodies or individuals. Once settled in a higher court the decision is 'on the record'; i.e. it is reported in the newspapers as a court report, finds its way into the law books (or it could be privately printed) and can be cited as conclusive or persuasive authority in courts all over the world.

6.3   Changing perception of 'cruelty'

As the years passed, however, agriculture evolved and developed; by 1960 intensive husbandry, especially of poultry and pigs, had become the method of choice for many farmers. Indeed, it is significant that the very word 'farmer' was being replaced by 'producer'. In 1964, the book *Animal Machines* was published (Harrison, 1964) and had an impact which is still being felt. Mrs Harrison argued that the methods and scale of modern husbandry systems had altered the living conditions of farm animals to such an extent that existing legislation was incapable of dealing with the complex ethical questions raised by the advent of intensive methods and was thus no longer adequate to protect their welfare. Her book struck a sympathetic chord, not merely with the public but among a number of farmers and agriculturalists, and resulted in the setting up of the Brambell Committee, which reported in 1965 (HMSO, 1965).

The Brambell Report was a very far-sighted document and correctly identified many of the questions, problems and implications which have subsequently arisen in the field of farm animal welfare regulation and legislation. For example, it drew attention to the importance of being able to define suffering and well-being and the key role of scientific research, to the need for acceptable and enforceable standards of accommodation and husbandry, and to economic implications such as the necessity for consistent regulations across free-trade areas.

As a direct consequence of the Brambell Report, the UK Government, after consulting with the agricultural industry and animal protection societies, passed the Agriculture (Miscellaneous Provisions) Act 1968. As a result of this Act, Codes of Recommendations for the Welfare of Livestock were produced for each of the important species of farm animal which, although not mandatory, involved a major change in thinking. Parliament agreed with the Brambell Committee that existing animal welfare legislation did not adequately safeguard farm animals. In the preamble to the Codes their status in relation to the law is set out and it is made clear that failing to observe a provision of the Code is not an offence in itself. However, if suffering occurs as a consequence, then the failure to observe the Code can be regarded as establishing the guilt of the accused. Brambell had distinguished three levels of suffering: 'discomfort' (recognized by poor condition, low activity and reduced appetite), 'stress' (tension or anxiety attributable to environmental causes) and 'pain'. In the Codes it states: 'Any person who causes unnecessary "pain" or unnecessary "distress" to any livestock . . . shall be guilty of an offence', i.e. two levels of suffering are recognized rather than three. Distress is not defined but presumably is intended to refer both to stress and the more severe and long-lasting forms of discomfort.

Another outcome of the 1968 Act was the setting up of a statutory committee, the Farm Animal Welfare Advisory Committee (FAWAC) to advise the Minister of Agriculture in matters of animal welfare. The Committee was (deliberately) broadly based, its members being drawn from the agricultural industry, from the scientific community, from the animal protection societies and from the public. In 1970, FAWAC published a two-part report on the working of the Codes which is interesting because it demonstrates a complete divergence in their conclusions between the members. The two parts were called the 'ethical' and the 'scientific' approach. Those supporting the ethical approach concluded: 'The crucial point relates to the limit which man, in a civilized society, is willing to take his exploitation of the animals he uses for food. Too much has been left in doubt and welfare needs have been placed second to considerations of productivity'. If, on the other hand, the scientific approach was followed, it was concluded that many of the recommendations in the Codes were more generous to the animal than was justifiable scientifically (MAFF, 1970).

The Codes have been amended on a number of occasions and the introduction to the most recent edition (1987) states: 'The Code embodies the latest scientific advice and the best current husbandry practices and takes account of five basic needs: freedom from thirst, hunger and malnutrition; appropriate comfort and shelter; the prevention or rapid diagnosis and treatment of injury, disease or infestation; freedom from fear; and freedom to display most normal patterns of behaviour'.

6.4 Animal welfare legislation in the continental countries of Europe

There are two major groupings of European countries which influence animal welfare legislation: the EC, with its 12 members, and the Council of Europe, which involves all the countries of Western Europe. Before looking at the progress which is being made towards harmonization of legislation in the EC, we shall consider the situation in a number of individual countries.

In some, legislation is very comprehensive and comparable to that in the UK, consisting of codified lists of actions which are prohibited, rather like the UK's 1911 Act. Norway, Sweden, Denmark, The Netherlands and Germany have detailed laws. In other countries general animal protection laws consist of little more than a few brief sentences which simply state that animals must not be ill-treated. Austria, Belgium, France, Italy and Luxembourg have adopted this more non-specific approach. It is interesting that the boundary line between the two approaches produces an approximate north/south division and it should be borne in mind that the legislation, which is extensive (W. T. Jackson, 1980; 1989) is undoubtedly enforced much more strictly in some countries than in others.

The most far-reaching legislation is the Tierschutzgesetz passed by the Federal German Parliament in 1972. It states that its basic principle is 'to protect the well-being of the animal. Without reasonable cause no one shall cause pain, suffering or injury to an animal'. The second part of the Act, which deals with the keeping of animals, says that a person who is keeping or looking after an animal:

– shall give the animal adequate food and care suitable for its species and must provide accommodation which takes account of its natural behaviour;
– shall not permanently so restrict the needs of an animal of that species for movement and exercise that the animal is exposed to avoidable pain, suffering or injury.

Denmark also has comprehensive legislation – for example, the Protection of Animals Act 1950, states that:

– animals must be properly treated and must not by neglect, overstrain or in any other way be subject to unnecessary suffering, and
– anyone keeping animals should see that they have sufficient and suitable food and drink, and that they are properly cared for in suitable accommodation.

This was the Act that was interpreted as prohibiting the keeping of hens in battery cages, until that particular provision was reversed in the 1970s.

The pattern, country by country, in general terms remains much the same, even though there are variations in detail. Luxembourg, for example, prohibits the housing of animals in such a manner that they suffer from lack of space in the stall or enclosure in which they are kept, from inadequate lighting and from lack of protection from the elements. In Italy, regulations prohibit the housing of veal calves in confined spaces. However, as in the UK with the 1911 Act, a great deal depends on how the terms are interpreted: when does suffering become unnecessary, how much is sufficient food and drink, how is suitable accommodation defined and what is confined space? Spain and Portugal have no specific measures as yet. Although Greece has no special law for the protection of animals kept for farming purposes, it will rely on the ratification of the European Convention.

In addition to laws drawn up with the specific aim of improving welfare, there is legislation framed for other purposes which may have effects on poultry husbandry systems and management, and thus, indirectly, influence welfare. Examples which fall into this category include the EC Egg Marketing Regulations which set out trading standards for eggs produced in different systems. In Norway a law has been passed which limits the number of birds which can be kept on a farm. Although its purpose is to encourage rural employment it will, by placing a ceiling on flock size, tend to discourage very intensive husbandry systems. Again, laws on environmental impact and pollution such as those being proposed in the Netherlands, may discourage large scale intensive systems by limiting the quantities of waste material which can be produced.

6.5   Council of Europe and the European Community

The aims of the Council of Europe, which comprises virtually all the nations of Western Europe, include the achievement of a greater unity between its member states by safeguarding the ideals and principles which are their common heritage through agreements and action in economic, social, cultural, scientific, legal and administrative matters.

One of the areas in which the Council has been active is animal welfare. Indeed, it has stated that: 'the humane treatment of animals is one of the hallmarks of Western civilization'. In 1976 it adopted the Convention on the Protection of Animals kept for Farming Purposes, which was concerned with the care, husbandry and housing of farm animals, especially those in intensive systems. Its recommendations are couched in general terms, but the drafting committee commented that they tried to lay down principles which are precise enough to avoid a completely free interpretation, but wide enough to allow for different requirements. Article 3, for example, states: 'Animals shall be housed and provided with food, water and care which – having regard for their species and to their degree of development, adaptation and domestication – is appropriate to their physiological and ethological needs, in accordance with established

experience and scientific knowledge'. Article 4 states: 'The freedom of movement appropriate to an animal, having regard to its species and in accordance
with established experience and scientific knowledge, shall not be restricted in
such a manner as to cause it unnecessary suffering or injury. Where an animal is
continuously tethered or confined it shall be given the space appropriate to its
physiological and ethological needs'. Article 5 deals with lighting, temperature,
humidity, air circulation, ventilation and other environmental conditions such as
gas concentration and noise intensity. Article 6 deals with the provision of food
and water and Article 7 with inspection – both of the condition and state of the
animal and of the technical equipment and systems.

The EC became a party to the Convention in 1978. Because the Convention
itself is very broad the Council of Europe has a Standing Committee with a
responsibility for elaborating more specific requirements and one of the first
areas in which it became active was that of poultry welfare. After five years of
negotiation an EC Directive was adopted in 1986 laying down minimum standards for the protection of hens in battery cages (Council Directive, 1986). By
January 1988 all newly-built cages had to provide:

– a minimum area of 450 cm^2 per bird and 10 cm of feeding trough per bird;
– a continuous length of drinking trough providing at least 10 cm per bird or if
nipple drinkers or drinking cups are used, at least two shall be within reach of
each cage;
– cage height of at least 40 cm over 65% of the cage area and nowhere less than
35 cm; and
– cage floors capable of supporting adequately each forward-facing claw and
not sloping more than 8°, unless constructed of other than rectangular wire
mesh.

These standards were to apply to all cages in January 1995. The Schedule to the
Directive makes further requirements.

Directives have to be translated into national legislation; in the case of the
UK, for example, this was done by adopting Statutory Instrument 1987 No.
2020 (HMSO, 1987). These regulations included the provisions set out in the
EC Directive and contained some additional ones, requiring extra space when
only one, two or three birds were kept in a cage (section 4.9). At the time of
writing there is a proposed amendment to SI 1987 No. 2020 which will require
that cages have fully opening fronts, in order to allow birds to be removed from
them with less risk of injury. Coupled with the main regulations was a schedule
setting out further requirements; these laid down specifications for cage design
and construction, for daily provision of adequate and nutritious food and water,
for good environmental standards including temperature and air quality, for a
suitable diurnal lighting pattern, for thorough daily inspection of the birds, for
prompt remedial action in the case of health or behavioural problems and for
the adequate functioning of automated equipment together with satisfactory
back-up systems in cases of failure. These regulations will in future be under

constant scrutiny; there will, for example be a 1993 Review of Directive 86/113/EEC.

In the case of Denmark a minimum of 600 cm^2/bird is required and in Germany 450 cm^2/bird up to 2 kg and 550 cm^2/bird over 2 kg. In the Republic of Ireland (Statutory Instrument 238, 1990), in addition to the cage regulations, requirements for non-cage systems are also laid down.

The impetus for framing EC Directives on welfare is, in general, directed more towards achieving uniformity of production standards than towards any ethical concern for animal suffering (C. Jackson, 1989). This impetus stems from the fact that countries unilaterally enforcing higher standards of welfare would in many cases increase production costs, thus placing their producers at an economic disadvantage compared to others with lower standards. Such differences would either act as a hindrance to free trade or could serve as a justification for limiting imports from countries with poorer welfare standards. Only states such as Switzerland which are outside economic blocs such as the EC or EFTA (European Free Trade Association) are able to establish their own independent regulations; they can ban imports from systems which fail to match Swiss welfare requirements. Nevertheless, even though the rationale for uniform welfare standards in the EC is an economic one, the influence of EC laws in the wider world is likely to increase. In order to continue trading with the EC, countries outside it will, in due course, have to match EC welfare standards in their own legislation.

This economic background is implicit in the Marketing Regulations set out in an annex to Commission Regulation EC 1943/85, which came into force in July 1985. It laid down criteria to be met by poultry enterprises producing non-cage eggs, which have to be sold under one of four possible labels: free range, semi-intensive, deep litter and perchery (or barn) eggs. The detailed criteria are set out in Table 4.1.

The north/south division in legislative approach is also reflected in when particular states ratified the European Convention on the Protection of Animals kept for Farming Purposes This may be linked to the proportion of the population engaged in agriculture, which on average is much smaller in the north of Europe than in the south (Politiek and Bakker, 1982). For example, the UK has 2%, Belgium, Luxembourg and West Germany 4%, The Netherlands, Sweden and Switzerland 5%, Norway and Denmark 8%, while France has 9%. All these countries had ratified the Convention by 1982. In the south, in contrast, Italy has 12%, Spain 17%, Portugal 26%, Greece 30% and Turkey 54%. In 1982, none of these had ratified the Convention. The only exception to the north/south division was the Republic of Ireland, which has 23% of its people engaged in agriculture and had not ratified the Convention. It may be, on the one hand, that pressure for increased animal protection comes primarily from a largely urbanized population detached from the practical problems of earning a living from animal husbandry, or on the other that where many people are engaged in agriculture their governments are unwilling to impose restrictions which affect their livelihood.

6.6   North America

In the USA, statutory legislation dates back to 1641 when the Massachusetts Bay Colony framed their first legal code. Clause number 92 read: 'No man shall exercise any tyranny or cruelty towards any brute creatures which are usually kept for man's use'. Their measure was far ahead of its time, for not until 1828 did one of the States of the Union pass an anti-cruelty law. The New York State Legislature in that year gave protection to horses, asses, cattle and sheep against malicious killing, maiming and wounding, cruel beating or torturing. Anyone committing these acts would be adjudged guilty of a misdemeanour. This example was followed in 1835 by Massachusetts and in 1838 by Connecticut and Wisconsin, with 20 other States following in the 1840s, 1850s and 1860s.

The American Society for the Prevention of Cruelty to Animals was formed in 1866, and its founder, Henry Bergh, was concerned that convictions were rarely, if ever, obtained under the existing legislation. Accordingly, he drew up Statutes for the State of New York which were passed in 1867 and subsequently used as a model by many other States. These statutes were much more specific and precise than previous laws, covering, as well as the more obvious cruelties, lack of provision of food or water, and transportation in a cruel or inhuman manner. They applied to 'any living creature'. Analysis of current laws shows that the public conscience generally agrees that animals have the right to: protection from cruel treatment, protection from abandonment, protection from poisoning and the provision of food, water and shelter (AWI, 1991).

Very few statutes relate specifically to farm animals, other than Federal laws on humane slaughter and transportation. There are no laws applying especially to poultry, other than State laws prohibiting the inhumane transport of poultry (AW1, 1991). Six States specify that crates for holding poultry shall conform to various requirements, often very modest. For example, Pennsylvania requires that live poultry shall not be stocked at more than 15 pounds per cubic foot; this is equivalent to about 248 kg per m^2. In Wisconsin it is unlawful to transport chickens in coops unless the coops are 13 in (33 cm) high inside.

In Canada, a Code of Practice for the Care and Handling of Poultry was published (Agriculture Canada, 1989) which, though closely based in its approach and format on that produced by the UK Ministry of Agriculture, Fisheries and Food, transcends it both in content and in detail. For many areas it provides sufficient information to serve as a basic handbook of good poultry husbandry.

6.7   Welfare societies

The general concern for the welfare of animals is reflected in the large numbers of societies and groups which have been set up in most countries. In the UK, for example, the first of these was the Royal Society for the Prevention of Cruelty to Animals (RSPCA), which can trace its origins back to 1835. It is a 'mainstream' animal protection society which is active on behalf of livestock, companion animals, experimental and laboratory animals and those in circuses and zoos. Its role includes both lobbying for a wide range of reforms and providing practical assistance for animals in any kind of difficulty, such as being abandoned, trapped or injured. This means that this society is involved, among many other interests, in arguing for improved housing and conditions for farm and zoo animals, calling for a ban on fox and stag hunting, pushing for a reduction in the number of experiments carried out on animals, and pressing for the compulsory registration of dogs.

Other societies tend to have special interests and concentrate their efforts on more specific issues. They include Compassion in World Farming, which campaigns for a ban on the export of live animals, the Farm and Food Society, which is concerned with intensive husbandry, the Farm Animal Care Trust which works for improvements in the welfare of farm animals and Chickens' Lib, with a similar role but concentrating on poultry welfare. There are even regional societies, such as the St Andrew Animal Fund, based in Edinburgh, which directs much of its energy towards reducing the number of scientific experiments carried out in Scotland. These societies are regarded as 'mainstream' in having relatively moderate aims and using legal methods. They are generally well organized and pursue their objectives through lobbying Members of Parliament, circulating literature to their own members and to the public, writing to the press and placing paid advertisements. Aware of the importance of influencing opinion at an early age they have also recently begun to produce information packs and video tapes directed towards school teachers and children. They may thus exert an influence out of all proportion to the number of their members.

On the fringe of the welfare movement there are other organizations, from non-violent activists such as hunt saboteurs to shadowy extremists in loosely organized groups such as the Animal Liberation Front (ALF) who, as well as 'releasing' laboratory or farm animals, engage in terrorist activities involving damage to property and even threats to human life.

6.8   Agricultural organizations

Booklets are produced by the National Farmers Union (NFU, 1990) and the British Egg Information Service (BEIS, 1990) which provide concise facts and summaries of the history, scope and present structure of the egg industry. Both present clear descriptions of the range of different production systems, briefly set out the advantages and disadvantages of each and emphasize the importance of safeguarding the hen's welfare. The NFU publication has a short section on poultry meat, dealing primarily with broilers, though ducks and geese are mentioned.

6.9   Public opinion

The welfare of domestic poultry has been a controversial issue in the UK at least since the early 1960s, when systems such as deep litter and free range were being phased out and replaced by cages. Public awareness of intensive husbandry methods was increased by the publication of *Animal Machines* (Harrison, 1964) and the Brambell Report (HMSO, 1965), and has been maintained by the activities of animal protection societies. Members of Parliament have stated that, over the years, the single issue about which constituents write to them most often is animal welfare.

Spokesmen for the agricultural industry have consistently argued that any system in which animals are physically healthy and in which production is good must, therefore, be a humane and satisfactory one. Such arguments, although not entirely without merit, are suggestive of special pleading, are based on narrow definitions of health and productivity, and tend to be framed from the viewpoint of the producer rather than from that of the animal. They have proved unconvincing as far as public opinion is concerned. As far back as 1968 a Gallup opinion poll showed that most people were in broad agreement with Brambell's conclusions and were unhappy with intensive husbandry systems which resulted in animals being closely confined. About 90% of people felt that farm animals should be able to turn around freely in their pens and that poultry should have enough room to spread their wings (Social Surveys, 1968). A substantial majority (79%) also considered that all livestock, including poultry, should have access to the open for some part of the day during fair weather. Such attitudes were not confined to the general public – a substantial proportion of farmers polled by Gallup held similar opinions.

These pro-animal views had changed very little by 1983, when National Opinion Polls (NOP, 1983) found that 90% would support a change in the law in order to prevent animals from being kept in conditions where they were unable to turn around, stretch their limbs or groom themselves. A similar proportion were in favour of introducing legislation to improve conditions for intensively housed animals. There were few differences in attitude across age groups or social classes, but women were more strongly in favour of changing the law than men.

A second survey, carried out at about the same time (MORI, 1983) yielded slightly different results. When people were asked about farming practices involving intensive housing or overcrowding, 47% responded that they should be banned, while 47% felt they were unavoidable. Again, more women than men were in favour of laws banning intensive husbandry practices. This time restrictive legislation was favoured slightly more by older people and by those in the skilled and unskilled working class (social classes C, D and E). Those with the least experience of agriculture favoured a ban more than those who had close contact with farming. The majority of respondents (63%) indicated a willingness to pay more for their food in order to ensure better treatment for animals. Housewives and the middle class (social classes A and B) were more likely to say they would pay more, whereas men and the young were less willing. The survey brought some good news for farmers – 91% felt that farmers treated their animals either well or very well.

Further evidence of distaste for the close confinement of animals came in a pilot study into the attitudes of people towards different housing systems for laying hens (Rogers *et al.*, 1989). There were three groups, each containing 20 people, categorized as welfarists, agriculturalists and general public, who completed a questionnaire after watching a video tape showing the salient features of six housing systems: battery cages, cages modified with perches, nestboxes and dust baths, perchery, deep litter, strawyard and free range. There were striking similarities between the groups as well as some differences. All apparently preferred less intensive systems, with free range given the highest rating and cages the lowest (Figure 6.1). The similarity in overall ratings given was despite the fact that groups listed different factors as important when deciding how to house hens. Economics was regarded as the most important factor by the agri-

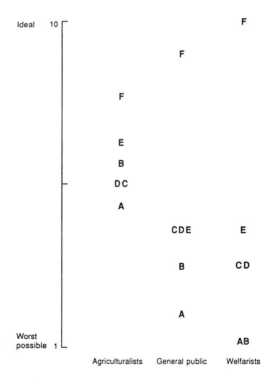

Figure 6.1. Overall ratings given to poultry housing systems by three categories of people after watching a video in which the six systems were shown. A, battery cages; B, modified cages; C, perchery; D, deep litter; E, covered strawyard; F, free range (Rogers *et al.*, 1989).

culturalists, whereas the welfarists and the general public placed the behaviour of the birds first. Agriculturalists tended to assess all the systems as better than welfarists did and with less spread between the best and the worst. The reactions of the general public were intermediate. Because this was a small-scale study the results should be viewed with caution but they do suggest that some underlying attitudes, such as a marked preference for less intensive methods of poultry husbandry, are firmly rooted in the population.

Public opinion has had its most far-reaching impact in Switzerland, where a referendum on housing systems for laying hens in 1978 resulted in cages being phased out by January 1992; less intensive approved housing methods will have to be used instead.

A UK Select Committee on Agriculture which reported in 1981 also concluded that cages should be phased out, but their recommendations were not acted upon. In 1988 the Swedish Parliament passed legislation which will require battery cages to be replaced by less intensive, probably aviary-type, systems by 1998, provided suitable systems can be developed by that date. The Dutch, too, are in the process of legislating for more extensive systems, which in due course will probably entail the banning of conventional cages.

6.10   Ethical background to animal welfare

Any consideration of public opinion and animal welfare is incomplete without taking into account the underlying philosophical values in which it is rooted. The question 'to what extent should humans exploit animals?' is, at heart, a moral one. As the surveys of attitudes have shown, the vast majority of people (at least in the UK) consider that cruelty to animals is wrong and this is, of course, the basis on which all animal protection legislation is founded. However, it could be argued that many people hold attitudes which are inconsistent and, in pure ethical terms, difficult to defend. If it is wrong to make animals suffer then perhaps animals should not be kept in any husbandry system, because all systems, even the most traditional ones, involve some suffering. It is true that animals in the wild may well suffer very much more, but this is not seen as altering the ethical argument, because animals in their natural environment are not regarded as being a human responsibility. One wholly consistent attitude would be an absolutist one, adopting veganism and avoiding all animal products. Some welfarists do hold such views, but the average person's position is one of compromise, eating meat, milk and eggs, wearing leather and wool, but also demanding that these products are produced with as little suffering to the animal as possible. This might be termed a 'minimalist' position; it is the basis on which legislation is framed and explains why the wording of animal protection laws rely so heavily on the concept of 'unnecessary' suffering. Problems arise, of course, when we ask 'how much suffering is necessary, or acceptable?' and whether we mean acceptable to the animal or to the consumer.

When we examine the question in this way it becomes clear that the concept of animal welfare has three strands – ethical, sociological and scientific. The ethical strand derives from people's moral values, their belief that life itself is of value, that animals can suffer, that if we exploit animals for our own purposes then we should, in turn, ensure that they live as contented lives as possible and that agricultural systems should not depart too far from 'natural' conditions. Moral values are, of course, fundamental beliefs and do not require justification. The sociological strand implies that in a pluralist society there are broadly ranging views, some firmly held for moralist reasons and others much less strongly so. Legislation and codes of practice must attempt to satisfy as wide a range of these views as possible, without offending strongly held beliefs. The scientific strand provides decision makers and legislators with the material to fashion rational regulations which will meet the needs of farmers to have workable systems and practices, without offending the moral values of the majority.

The animal-centred view, in which the reactions and 'feelings' of animals themselves are regarded as paramount in deciding whether or not they are suffering, was foreshadowed in the Brambell Committee's Report (HMSO, 1965). Such an approach is held by most, if not all, applied ethologists working in the field of animal welfare research and meets this scientific demand because it concentrates on the effects of intensive agriculture on the animals themselves. The only evidence which can modify people's fundamental beliefs in this area is that which is relevant to whether or not animals are contented in a particular system or under a particular procedure. Thus, evidence which relates to animals' feelings is likely to be more compelling than that which relates merely to their physical condition.

6.11   Future developments in legislation

Some of those involved with the development and application of welfare legislation have recently been expressing dissatisfaction with the way that legislation has been framed in the past (Everton, 1989). It is considered that laws have been expressed in too negative a way: their approach has been to set limits on what can be done to animals by banning or discouraging practices or systems which are likely to result in suffering. Such an approach may have resulted in a reduction or elimination of the grosser forms of abuse but has had less effect in raising general standards. It is now felt that a better approach, which would be more effective in improving the quality of life for the average animal, would be to legislate for 'positive well-being'. This would require that all the needs of animals should be identified and that legislation should be framed placing a general duty on farmers, breeders, producers, transporters and slaughterers to take specified measures which meet these needs in full (Everton, 1989). Such a duty, for example, might require persons having livestock under their control to provide *ad libitum* fresh water, food of a quantity and quality sufficient to maintain the animals in full health and vigour, accommodation designed to enable livestock to stretch their limbs and move around without constraint, provide them with a dry, comfortable resting area, and so on. In the case of the UK this would mean that measures which are at present contained in Codes of Recommendations would become legal requirements. It would be necessary for animal scientists to carry out research which enabled the needs of livestock to be accurately defined. Implicit in the general requirement to promote well-being would also be a duty to protect livestock from avoidable suffering, which might be defined as pain, disease, distress, or acute or prolonged discomfort.

Welfarists and livestock industry alike agree that first-class stockmanship is the key to satisfactory welfare. In the UK certain establishments concerned with animals, such as riding schools, boarding kennels and pet shops already have to be approved and licensed. It is being increasingly argued that producers and farmers concerned with livestock should also require to be licensed in the same sort of way. One of the conditions of approval would be that they and their staff would have to demonstrate competence, initially perhaps on the basis of adequate experience, but eventually requiring evidence of formal training in stockmanship.

II

Behaviour, Management and Welfare

Part II presents a cross-sectional view of poultry production systems by showing how birds behave in different systems and considers the implications for management and welfare. After the first general chapter, it covers key aspects of behaviour and their interaction with the environment, opportunities for using them to best advantage and problems which may arise.

7

Control of behaviour

7.1 Summary

- Behaviour is partly under genetic control, so it is affected by natural selection and, in domestic animals, by artificial selection. Some changes under domestication have been due to conscious selection, but others are chance effects or due to linkage with other characteristics.
- Genetic effects interact with the environment in producing the final form of behaviour. Certain aspects of behaviour, such as nesting, are relatively 'pre-programmed', but their orientation and detailed expression depend on circumstances. For other types of behaviour, inherited genes allow flexibility and learning is more important.
- Poultry show simple forms of learning such as habituation and more complex forms such as conditioning and imprinting. These can be utilized both by changing the environment to influence behaviour indirectly and by adopting management practices to modify behaviour directly.
- External and internal stimuli affect the nervous system and physiology of animals and consequently influence behaviour. The stage in the processing of such information which is perceived by the animal is called motivation.
- Measurements of motivation indicate that for certain behaviour patterns it increases with time and inability to perform such behaviour results in frustration. Other behaviour is only motivated to restore some imbalance, for example, physiological. Animals may therefore have both physiological and behavioural needs, with different implications for housing and husbandry.

7.2 Evolution of behaviour

Behaviour, like all the other characteristics of organisms, has evolved through natural selection. Under natural conditions, those individuals whose behaviour

105

equips them best survive and leave the most offspring, which tend to inherit their parents' behaviour. Jungle fowl evolved in the rainforest of Southeast Asia and their behaviour patterns are those typical of ground-dwelling birds. They spend much of their time in cover and show appropriate species-specific protective responses to predators – remaining silent, crouching and freezing to overhead predators, while calling and running or flying away from ground predators.

The process of domestication can be regarded as a special, often accelerated, form of evolution and, together with the predictable effects on production traits such as more rapid growth rate, greater body size and increased egg output, has resulted in characteristic changes in behaviour. It is debatable, though, to what extent differences in behaviour between domesticated poultry and their ancestral forms are due to deliberate selection. Some patterns which first evolved in jungle fowl have remained almost unchanged in modern hens (e.g. nesting behaviour and anti-predator responses), presumably because they were widespread and stable in the genotype and there has been no selection against them. Other behaviour patterns have been strongly selected against and have almost disappeared from modern hybrids. An example is broodiness, which was inadvertently selected against as a consequence of selecting egg-laying hybrids on the basis of high egg number (section 10.7). Individuals which became broody during the test period would have laid fewer eggs and thus would have been withdrawn from the breeding population. Breeds such as bantams, which have not undergone intensive selection for egg output but instead have been kept for their appearance or their plumage, have retained broody characteristics.

In an elegant series of experiments on ducks, wild-type mallards and domestic Aylesbury ducks were hatched and reared under identical conditions so that any variation between the ancestral and the domestic type could be attributed to genetic rather than environmental causes (Desforges and Wood-Gush, 1975a, b). There was less aggression and more tolerance towards other individuals in the domestic ducks, their flight distance was much less when humans approached and they habituated more quickly to novel stimuli (Figure 7.1). They were also more willing to eat unfamiliar foods. Changes like these, which are representative of a broad range of domesticated species, are unlikely to be due to conscious selection by man, even though they are advantageous as far as management and feeding are concerned. Instead, they probably occurred because individuals showing these traits were best suited to the husbandry, dietary and housing conditions of domestication and left more offspring than those which were less well adapted.

In a few cases there has been selection for a particular behaviour. In the early days of domestication it is likely that fowl were kept mainly for cock-fighting, so cockerels were carefully selected for maximum aggression, only those which had won contests being retained for breeding. Modern cockerels crow much more frequently than jungle fowl; this is believed to be a relic of their involvement in religious ceremonies such as those of Zoroastrianism, in which crowing

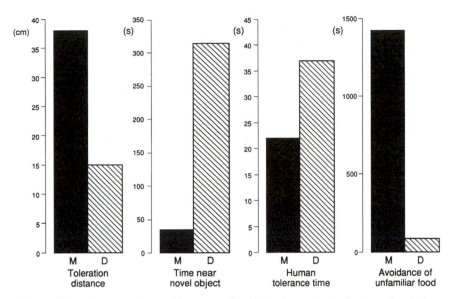

Figure 7.1. Compared to wild-type mallard (M), domestic ducks (D) tolerated other individuals nearer to them without aggression, spent more time close to a novel object, took longer to move away from a human being when released and fed sooner when food was presented in an unfamiliar container (Desforges and Wood-Gush, 1975a, b).

cocks took an important role and were, presumably, selected on the basis of loud, long and frequent crowing (Wood-Gush, 1959a).

Most selection on specifically behavioural traits has been on an experimental basis and examples are given in the appropriate chapters. It should be said, though, that there is increasing interest in such studies. The importance of a genetic component in behaviour means that certain traits which are today regarded as undesirable could be reduced or eliminated by selection – possible examples would be fearful or panic responses, cannibalism, feather pecking and inter-individual aggression in laying flocks.

There are substantial differences between strains in the incidence of pre-laying agitation and pacing often seen when hens lay in cages (Mills *et al.*, 1985a), as Figure 7.2 shows. The various elements of this behaviour are under partial genetic control and so they could, at least in principle, be selected against (Braastad and Katle, 1989; Heil *et al.*, 1990). Whether this would be beneficial for welfare is, however, an open question. The selection might be only against the outward display of frustration, in which case the underlying distress might become manifest in some other way.

There is also interest in controlling for other aspects of welfare during selection. As one example, selection for high growth rate in broilers has led to an

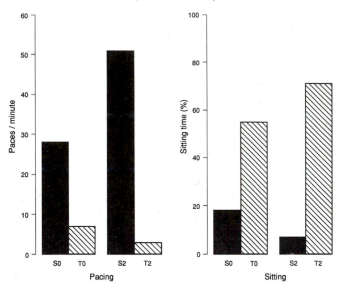

Figure 7.2.   Before genetic selection, during the 10 min period before laying, S-line (S0) hens paced more than T-line (T0) hens and spent less time sitting. After two generations of bidirectional selection, for pacing in the S-line and for sitting in the T-line, these differences were exaggerated (Mills *et al.*, 1985a).

increase in leg disorders. Some of these have a strong heritable component and can be eliminated without much effect on growth (Sorensen, 1989).

The fact that selection for certain traits often produces changes in others has had effects on behaviour. The most obvious of these is the increased placidity in strains selected for meat production, especially broilers. The same effect accounts for differences in flightiness between medium-weight hens like the Rhode Island Red, originally selected as a dual-purpose breed for both meat and egg production and the more nervous light-weight breeds such as White Leghorns, selected only for eggs. It is reasonable that higher growth rate should be associated with a calmer disposition, but the actual mechanism is not known.

7.3   Interaction of genome and environment

Although genetic factors play a major role, perhaps more so in birds than in mammals, behaviour observed in an individual animal is in part inherited and in part learnt. Birds possess a relatively large proportion of species-specific behaviour patterns – 'innate' or 'pre-programmed' – which are subsequently modified by interaction with the environment. A good example of this interaction is provided by maternal imprinting: chicks have an innate tendency to imprint on the first moving object they see and will thus treat an artificial object as a surrogate parent. Thus, the primary tendency to follow is inherited, but they learn what to follow. In fact, it is more complex than this – there are also constraints, which are inherited, governing imprinting. If the object is about the size of the mother bird they are more likely to imprint than if it is much smaller. If the object emits calling sounds, similar to those of a hen, then they are more likely to imprint than if it is silent. If it is patterned or moves they are more likely to imprint than if it is plain or stationary. There are interesting between-species differences in ducks which are clearly adaptive: mallard, which are ground nesting, show visual imprinting whereas wood duck, which are hole nesters in trees and thus rear the young in the dark, show auditory imprinting (Klopfer, 1959).

Other behaviour also shows this pattern of an innate base which is modified appropriately by environmental factors. Soon after they have hatched, 1-day-old chicks show an innate tendency to peck at a wide range of stimuli around them. At this stage they will peck equally at grains of sand or at particles of food (Hogan, 1971). However, as time passes the pecking at sand wanes while the pecking at food strengthens in response to the positive nutritive feedback.

Nesting behaviour appears as a fully organized collection of behaviour patterns, without any discernible developmental stage, on the very first occasion when a point-of-lay pullet lays an egg. However, the precise nature of its expression thereafter is greatly modified by the surroundings in which the bird finds herself. Birds with access to adequate nest sites display a full repertoire. In cages the investigative phase may be replaced by a prolonged period of stereotyped pacing or escape behaviour, while the final sitting phase may be almost entirely absent (section 11.6).

A final example is provided by the fearfulness which many birds display towards humans. The level of this fear is very high in jungle fowl but selection during the course of domestication has considerably reduced it in modern strains. It can be reduced still further by environmental factors (Figure 7.3), such as the exposure and regular handling of young chicks by humans (Jones, 1987b) and their exposure to enriched environments (Jones, 1989).

7.4   Development of behaviour

Learning can be defined very broadly as 'internal change causing adaptive modifications in behaviour as a result of experience' (Thorpe, 1951). The simplest form is habituation, in which a response, after a number of repetitions, gradually wanes over time. An example of this is seen when chicks are frightened by an object passing over them, whereupon they show an anti-predator response; they stop whatever they are doing, crouch and freeze. If the action is repeated a number of times in rapid succession without any unpleasant consequences for the chicks, the duration of freezing becomes shorter and shorter, until eventually they ignore the stimulus altogether. This is an innate response which is modified by the bird's experience – but the response is restored by the

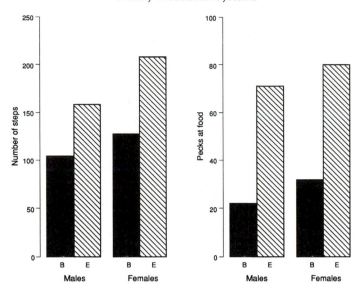

Figure 7.3.   Male and female domestic chicks reared in barren (B) rather than an enriched (E) environment spend less time moving around and peck less at food when they are placed in a novel enclosure for 15 min (Jones, 1982).

passage of time or if there is any appreciable change in the nature of the stimulus. Habituation is a useful reaction, it allows animals to adapt quickly to prominent but non-noxious features of the environment which initially caused them disturbance. This effect is probably even more important in commercial conditions than in the natural situation.

Classical conditioning is the type of learning first described by Pavlov. Dogs normally salivate in anticipation if presented with food such as meat. What Pavlov found was that dogs would learn to salivate in response to a stimulus such as the ticking of a metronome, provided the metronome was activated on a number of occasions immediately before a piece of meat was placed in the dog's mouth. In the same way hens will learn to respond to a previously neutral stimulus if it occurs sufficiently often in association with another stimulus, which itself causes a predictable response. For example, in one experiment light hybrid hens were given a fright by suddenly inflating a balloon close to them and usually showed a predictable escape response: they jumped up, wing flapped and squawked. If a lamp was switched on a few seconds before the balloon was inflated, the birds soon became disturbed by the lamp on its own and often showed the escape response even if the balloon was not inflated (Rutter and Duncan, 1992). This kind of associative learning is obviously adaptive under natural circumstances: animals which learn to avoid stimuli associated with the presence of predators are less likely to be caught and preyed upon.

Another form of learning is operant conditioning, in which animals learn to carry out behaviour in order to obtain food, water or some other desirable consequence. Although closely linked with the name of B.F. Skinner, it rests on the principle first propounded by Thorndike, the 'law of effect'. Behaviour followed by a pleasant outcome is more likely to occur again, whereas that with neutral or unpleasant consequences is less likely to be performed in future. The consequence of the operant behaviour is described as 'reinforcement' and can be positive or negative. Thus hens can be trained to peck at a small disc to operate a feeding device which gives them access to food and this is, of course, positive reinforcement. If, however, pecking the disc instead has unpleasant consequences, such as exposure to a stimulus they find frightening, they quickly learn to stop pecking. This is punishment. This type of learning again is adaptive – chicks which at first will peck at a wide range of small objects continue to peck at food particles because eating them has pleasant consequences, whereas they soon cease pecking at sand because its consumption does not provide them with positive reinforcement.

Imprinting is a special form of learning during which a newly hatched chick learns the characteristics of a parent (or, under experimental conditions, a surrogate parent, see previous section); it occurs during a sensitive period which lasts 24 to 36 h after hatching. The imprinted chick is able to learn the specific characteristics of relatively similar objects: it has been shown that domestic chicks exposed to a human being for 15 min are able to distinguish between the person on whom they have imprinted and a stranger (Gray and Howard, 1957). Under natural conditions imprinting is adaptive because the chick bonds to its mother and follows her closely, thus reducing the risk of separation during a vulnerable stage of its life. Poultry under commercial conditions are taken from the incubator trays shortly after hatching, swiftly handled and sorted, placed in boxes and transported to a rearing unit, where they are brooded in large groups. Thus they do not see their mother and cannot imprint on her, while any visual contact with humans is of brief duration and of the wrong nature to induce imprinting. If they imprint at all it will be on other chicks in their group, because they are likely to be the only moving objects which they see for any length of time during the sensitive period. This imprinting may help to explain why removing young chicks and isolating them from their group mates causes considerable distress, generally manifested as prolonged peeping calls.

7.5 Modification of behaviour

Our increased knowledge of poultry behaviour and the factors which influence it can, in some cases, now serve a practical purpose. Evidence is accumulating that the early experience of birds may have a considerable effect on their subsequent behaviour.

Enriching the environment of both domestic chicks and quail chicks by exposing them to a range of novel objects or stimuli makes them more able to resist disturbance and stress later in life – in particular, they seem to be less fearful (Jones, 1987b; Jones *et al.*, 1991). This was put to practical effect by one broiler producer, who adopted the practice of walking regularly through his flock of young chickens while loudly banging a metal can with an iron bar. This habituated them to disturbance, making them more placid and easier to catch and place in crates when the time came to empty the house (I.J.H. Duncan, personal communication).

In a similar fashion, giving pullets access to perches during the rearing period reduces the proportion of floor eggs when they are subsequently housed in multi-level systems such as percheries or aviaries (section 11.5). This is presumably because perch-reared birds become accustomed to moving around in three dimensions and thus are able to reach raised nest boxes without difficulty.

Environmental factors are frequently important in influencing undesirable behaviour. The control of pecking and cannibalism still poses major problems but a number of influential factors, such as complexity of the environment, light intensity, group size, stocking density and disturbance have been identified (Hughes and Duncan, 1972) as Figure 7.4 shows. Thus the conditions under which it is likely to occur can be predicted, even though the precise factors which trigger a particular outbreak often cannot be identified. Because there are large differences between strains, genetic selection is also likely to be effective in modifying the behaviour.

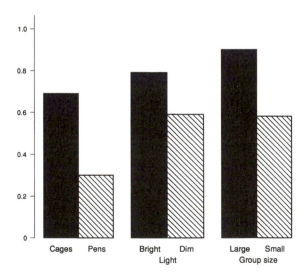

Figure 7.4. Pecking damage (on a scale from 0–4) was greater in cages than in pens, when light intensity was bright rather than dim and when birds were caged in large rather than small groups (Hughes and Duncan, 1972).

7.6 Physiological factors

One of the important functions of behaviour is to help maintain the constancy of an animal's internal environment – a good example of this is ingestive behaviour. The control of feeding behaviour is still not fully understood but it is clear that parts of the hypothalamus, an area at the base of the brain, are concerned with the initiation and termination of feeding. Activity in the lateral hypothalamus increases if animals are kept without food for long periods, in response to physiological changes such as an empty gut, mobilization of glycogen stores from the liver or a reduction in circulating metabolites such as glucose. This part of the brain could thus be regarded as a 'feeding centre' and if it is artificially stimulated the animal begins to eat. The lower-central part of the hypothalamus, on the other hand, is a 'satiety centre' and animals stop feeding when it is active; it responds to stimuli such as a full gut.

A number of hormones have important roles in the control of behaviour, especially that connected with reproduction, aggression and responses to stress. This control is very complex, often involving four stages. Substances called releasing factors are secreted by the hypothalamus into specialized blood vessels which supply a gland below the brain called the pituitary. There they in turn release hormones into the bloodstream. An example of one of these hormones is adrenocorticotrophic hormone (ACTH) which, in the third stage of the process, stimulates the adrenal glands to produce another hormone, the steroid called corticosterone (Figure 7.5). This finally binds to special receptors in various parts of the body and the central nervous system to affect their activity. Corticosterone, for example, acts on the brain to influence behaviour by changes in perception.

Much of our knowledge on the effects of hormones has come through experiments in which they are isolated or synthesized and injected into animals; this may give an indication, for example, of whether they have some involvement in a specific behaviour. However, the results can provide only a partial picture for a variety of reasons: the injected hormone may be present at concentrations very different to the normal levels; it may be converted in the body to other hormones which have different effects; it may also, through negative feedback, inhibit the normal secretion of related hormones by the animal's own glands. Thus, although testosterone injected into chicks results in precocious male sexual behaviour, such as mounting, treading and crowing, it is probably a metabolite of testosterone, oestradiol, produced within the brain, which is most important in maintaining the behaviour. Experiments investigating such complex processes therefore need very careful design to identify the precise mechanisms involved. In addition, this type of research raises ethical questions in that it can be justified only if the benefits likely to flow from it exceed the costs to the animals involved. All such experiments in the UK are controlled by the Home Office, which licenses them only when satisfied that the cost–benefit analysis is a favourable one.

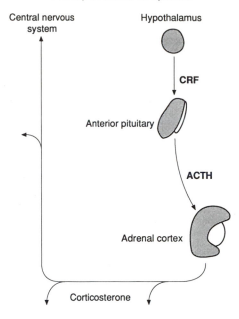

Figure 7.5.   Hormonal control of stress response. Corticotrophin releasing factor (CRF) is secreted by the hypothalamus and causes the anterior pituitary to release adrenocorticotrophic hormone (ACTH) into the circulation. ACTH stimulates the adrenal cortex to produce corticosterone, which among other actions can influence the activity of the brain.

Pre-laying behaviour is under hormonal control. Wood-Gush and Gilbert (1975) have shown that oestrogen and progesterone released from the ovary around the time of ovulation appear to initiate the sequence of pre-laying behaviour which results in nest-building and is terminated by the laying of the egg.

Similarly, incubation and broody behaviour, which are well developed in bantams, are closely related to the concentrations of circulating plasma hormones. As hens come towards the end of a cycle of laying, the level of luteinizing hormone (LH) in their blood begins to fall while that of prolactin rises and their tendency to sit tightly on their eggs increases. LH concentrations then remain low and prolactin high for as long as broodiness persists (Sharp *et al.*, 1988).

Circulating hormone concentrations may also play a role in the initiation of feather pecking and cannibalism in the domestic fowl: these behaviour patterns are commonly observed in females and rarely, if at all, in males. In females they often peak at around sexual maturity and their incidence can be manipulated by the injection of sex hormones. Thus in immature females their incidence is increased by the injection of progesterone, especially in combination with oestradiol (Figure 7.6), while it is reduced by testosterone (Hughes, 1973).

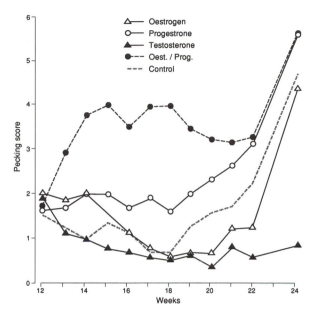

Figure 7.6. Pecking damage (on a scale of 0 to 6) in pullets is increased slightly by administering progesterone, and greatly by oestrogen and progesterone together. It is reduced by testosterone, which completely prevents the 'normal' point-of-lay rise (Hughes, 1973).

7.7 Motivation

An early, simple explanation for the occurrence of behaviour, proposed for instance by Descartes in the seventeenth century, was that it was an automatic response to an external stimulus. However, it became clear that this simplistic approach could not account for the complexity of behaviour patterns; for example, when food was offered to an animal it would sometimes eat it avidly, whereas on other occasions it would show little interest and turn away. It was, therefore, necessary to postulate the existence of some variable internal condition within the animal. This condition is called motivation, which has been defined as 'the causal state which is generated by all the stimuli (internal and external) which impinge upon an animal' (Toates, 1986). As already described, internal factors such as physiological deficits or hormone concentrations can exert a major influence on the expression of behaviour and it must be emphasized that motivation refers to the strength of the tendency to engage in behaviour when taking into account both the animal's internal state and the relevant external factors.

Attempts have been made to represent the relationships between internal and external factors and their effects on behavioural output in the form of models of motivation. These models fall into two main classes: those which incorporate an idea of some sort of energy steadily building up and acting as a force to drive behaviour and those in which behaviour is visualized as an agent for restoring homeostasis in a system. The relative value of these two types of model is discussed in the next section. An example of the first type is that put forward by Lorenz (Manning, 1967); it is often termed a 'psychohydraulic' model because it takes the form of a reservoir in which fluid (representing the energy specific to a particular type of behaviour) accumulates over time (Figure 7.7). As the level builds up the increased pressure opens a valve and the fluid is discharged. The valve can also be opened by weights, which represent releasing stimuli. The flow of fluid is an analogy for the expression of behaviour.

An important deficiency in Lorenz's model is the minimal contribution of feedback. This is rectified in the second type which is the homeostatic model (Toates, 1986). In this, the role of behaviour is to restore an animal which is hungry, or thirsty, or too hot or cold, or short of sleep, or deprived in any other way, to its normal state. The actual value of some variable is compared with a set point or ideal value (Figure 7.8). If there is a mismatch this is registered by the comparator or error detector, which activates the motivational system. The appropriate behaviour is then initiated and shifts the physiological system concerned in the direction of its desired state. The actual value is again compared with the set point and, once they match, motivation is switched off and the behaviour ceases. Of course, this basic model must be a major simplification of what is actually going on. For example, a thirsty animal stops drinking long before an appreciable amount of water has been absorbed from the gut and before the behaviour can have had any effect in restoring physiological variables such as plasma concentrations or tissue water content to normal. The system must therefore be taking into account a great deal of additional information, such as effects of the water on the mouth and throat, swallowing movements, activity of the oesophagus, distension of the stomach and so on and then acting on a predictive basis. The behaviour stops as soon as sufficient water has been drunk to restore the deficit, even though a considerable further period has to elapse before the water is fully absorbed, the actual values come to coincide with ideal values and the mismatch is corrected.

The simple homeostatic model has been extended by Baxter (1983), in part to deal with this kind of difficulty; once behaviour has altered the animal's internal or external environment these functional consequences terminate the behaviour, both through the perception of external stimuli and by a direct inhibitory effect on the level of motivation itself. This negative feedback loop, by acting directly on the level of motivation, allows a much quicker response to the functional consequences.

7.8   Behavioural needs

Different models of motivation have different implications for animal welfare.

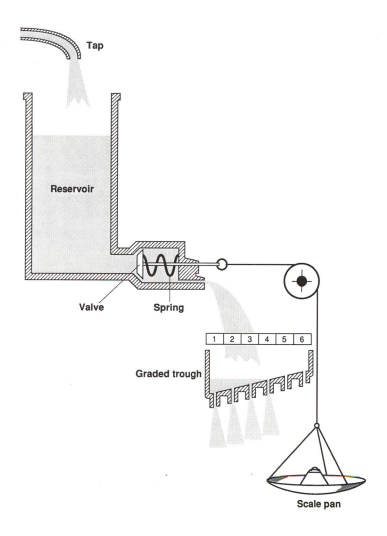

Figure 7.7. 'Psychohydraulic' model of motivation proposed by Lorenz (after Manning, 1967). Motivation is represented by water, which flows from the tap to fill the reservoir. The outflow represents the motor activity of behaviour, which is held in check by a valve and spring. The valve can be opened in two ways. Weights on the scale pan represent releasing stimuli and, if they exert enough force, will open the valve and release water, thus eliciting behaviour. Alternatively, as water accumulates in the reservoir its pressure will eventually become strong enough to open the valve. The graded trough shows how the behaviour elicited is related to the rate of flow of water – when more water is released the outputs of higher intensity behaviour increase.

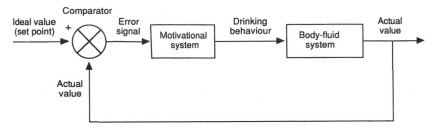

Figure 7.8.   Simple homeostatic model of motivation (Toates, 1986). The system contains a representation of an ideal state (or set point). The comparator measures the actual value at any time against the set point and, if any difference is detected, an error signal is produced. This activates the motivational system and, in the example given, drinking behaviour would be initiated, which would influence the body fluid system. The actual value would thus come closer to the ideal value, and the comparator would switch off motivation.

For example, in the UK the Brambell Committee considered the welfare of hens in battery cages (HMSO, 1965). In coming to their conclusion that 'much of the ingrained behaviour pattern is frustrated by caging' they leant heavily on the approach implicit in Lorenz's model. W.H. Thorpe, in his appendix to the Report, stated that 'a very large part of animal behaviour is basically determined by innate abilities, proclivities and dispositions' and went on to argue that 'we must draw the line at conditions which completely suppress all or nearly all the natural, instinctive urges and behaviour patterns'. This approach implies that if the tendency to perform behaviour builds up over time and the animal is unable to carry out the behaviour then it is likely to become frustrated and to suffer. Animals are thus said to have 'behavioural needs'. In fact, despite the probable restriction on behavioural expression in intensive systems, the Brambell Committee nevertheless concluded that well-designed, spacious cages were, on balance, the best method of keeping laying hens. This was because it considered that welfare would be seriously compromised in extensive systems such as free range through outbreaks of pecking and cannibalism.

More recent research on behavioural needs has been directed primarily at establishing to what extent the concept is valid. Wood-Gush (1973) pointed out that it was strongly dependent on which model of motivation was accepted. The Lorenzian model does indeed imply that motivation to perform a behaviour pattern will build up until, in a constraining environment, it is expressed in incomplete form, or as vacuum behaviour, or as a displacement activity. The homeostatic model, on the other hand, implies that behaviour is motivated only when the animal's physiological and psychological condition deviates from its normal state. If the homeostatic model is correct, provided all of an animal's basic needs are met it is unnecessary for any behaviour to occur (Baxter, 1983).

Neither model appears to be all-embracing. Lorenz's model fits some behaviour patterns well, whereas others are better accounted for by the homeostatic model (Toates and Jensen, 1990). Nesting behaviour, for example, fits the psychohydraulic model well. As the time for oviposition approaches birds first show evidence of increased activity, which becomes directed towards possible nest sites. They investigate these, often on a relatively systematic basis until one is chosen. Nest building behaviour is performed and the bird finally settles, sits and lays. If they are denied access to an acceptable nest site, as for example in a small cage, they may show characteristic pre-laying agitation and pacing, which has been interpreted as evidence of frustration (Duncan, 1970).

Other behaviour patterns, too, have at least some characteristics which fit the Lorenzian model: there is evidence that the tendency for hens to dust bathe increases with the passage of time (Vestergaard, 1980). Similarly, Nicol (1987b) found that the frequency of behaviour patterns such as wing stretching and flapping showed a rebound above their normal incidence after hens which had been confined for a period in small cages were moved into larger ones. She concluded that 'some continuous change occurs in the animal's state during the restrictive period'.

One of the issues relevant to behavioural need, which is at present under active research investigation, is the degree to which different behaviour patterns of poultry are stimulus bound, i.e. whether 'out-of-sight is out-of-mind'. As an example, if dust bathing were wholly stimulus bound, this would mean that it was motivated only by seeing dust, so that birds kept in cages would not be motivated to dust bathe, because they would never be stimulated by the sight of a suitable substrate. In this case, the question of behavioural need would not arise. In fact, hens housed in cages do show vacuum dust bathing, so stimulus binding for this behaviour is not complete.

7.9 Measuring motivation and behavioural need

It is becoming increasingly recognized (Dawkins, 1980; Duncan and Petherick, 1989; van Rooijen, 1985) that only by investigating the feelings of animals is it possible to safeguard their welfare fully. The first approaches in this direction involved examining hens' preferences for different aspects of their environment ('asking the animal') and it was shown that they chose cage floors with a small mesh size which gave more effective support for their feet (Figure 7.9) (Hughes and Black, 1973), chose a large rather than a small cage (Dawkins, 1981; Hughes, 1975a), chose enclosures which contained litter substrate rather than having a mesh floor (Hughes, 1976) and preferred to feed next to familiar cage mates rather than next to strange birds (Hughes, 1977).

Preference techniques also allow birds to be offered a choice between conditions which are both desirable. Hens showed a strong preference for large cages as opposed to small cages when both had a mesh floor and for litter rather than mesh when the cages were the same size, but when obliged to choose between a small cage with litter and a large cage with a wire floor they

Figure 7.9.   Hens prefer cage floors made of mesh with closely-spaced wires rather than with larger spaces, probably because a small mesh gives better support for their feet (Hughes and Black, 1973).

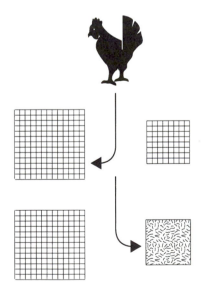

Figure 7.10.   Hens choose large rather than small cages, but prefer a small cage with litter to a large cage with a wire floor (Dawkins, 1981).

preferred the former (Figure 7.10). This suggests that they perceived litter as more important than space, at least in the context of that particular experiment (Dawkins, 1981).

The most reliable assessment of motivation is obtained when several different variables all indicate its existence – in the case of hunger, for example, the amount eaten, the rate of eating, the willingness to tolerate an unpalatable substance such as quinine in the food and the amount of work which an animal will carry out in order to obtain food. This last method provides an accurate measurement of motivation when birds are tested in Skinner boxes and have to peck at a disc in order to obtain a pellet of food. In the same way a number of separate tests have shown that access to a nest is important to hens: they are prepared to push through a weighted gate, pass through obstacles such as air blasts or pools of water and walk long distances in order to reach a suitable nest site (Duncan and Hughes, 1988).

These techniques are useful for measuring motivation but do not assess behavioural need. In order to do this the animal has to be provided with the goal

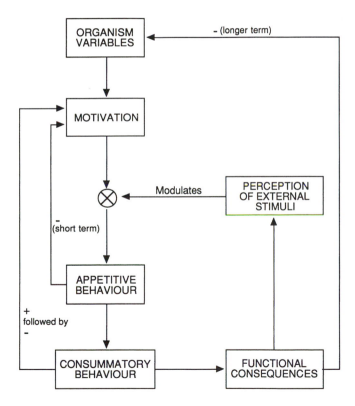

Figure 7.11.   Model of motivation proposed by Hughes and Duncan (1988).

and then observed to see whether it still carries out the behaviour. This was done by Hughes *et al.* (1989) who presented hens with pre-formed nests, which they had previously constructed themselves and found that they carried out just as much pre-laying behaviour as when they were given plain litter. If the behaviour is still seen when the goal is present it strongly suggests the existence of behavioural need.

The model of motivation developed by Hughes and Duncan (1988) gives a central role to behaviour, explaining why it may be important to allow animals to perform behaviour, rather than merely providing them with an environment which meets their basic nutritional and physiological requirements. It emphasizes the distinction between appetitive and consummatory behaviour, shows how behaviour at first has a positive feedback effect so that once behaviour has begun the animal becomes initially more inclined to continue and also makes it clear that motivation is eventually switched off both by the performance of behaviour as well as by its functional consequences (Figure 7.11). In the case of feeding, for example, hunger will be satisfied both by the action of eating as well as by the consumption of food. The increased expression of behaviour seen in alternative housing systems is obviously consistent with the conclusions drawn from the Hughes/Duncan model (Hughes and Duncan, 1988).

The development and control of behaviour is thus an extremely complex topic, but an increased knowledge of it is important, especially with the beginnings of a move to systems which give the opportunity for increased behavioural expression and the possibility of behavioural problems. Producers and stockworkers will need to understand what is going on in order to control it.

8

Feeding and drinking

8.1 Summary

- Young birds peck at small particles and shiny surfaces and gradually learn to distinguish food and water. In natural conditions they also learn from their mother, which strengthens species-specific diets: these tend to be very varied in galliforms, less so in other species such as geese.
- Palatability of food involves texture, colour, taste and smell and birds' familiarity with these. Their reaction to such features may also reflect the nutritional content: poultry can select a balanced diet from varied components and compensate for a number of specific deficiencies.
- Birds tend to feed together where possible, because of diurnal rhythms and social effects. Timing of feeding and time spent feeding are also influenced by the nature of the food and details of the light regime.
- Time spent foraging varies between systems. In systems where foraging is restricted to manipulation of the food itself, this often results in food wastage and may cause frustration or encourage feather pecking. Food intake also varies, affected by temperature and activity.
- Drinking is generally associated with feeding. Supervision of water supply is important, especially when birds are given a new water source. Drinking may also be an indicator of stress: some birds drink excessively in barren conditions or as a response to food restriction.

8.2 Natural behaviour and diet

Under natural conditions the jungle fowl's diet is a very mixed one – seeds, fruits, herbage and invertebrates. It browses on herbage and forages by scratching at the

123

ground, exposing small food items which are pecked up. The scratching and ground pecking can be regarded as appetitive behaviour, while the picking up and swallowing of the food is consummatory behaviour. Feral fowls on a tropical island ate berries and seeds, figs, leaves, isopods, insects and carrion (McBride *et al.*, 1969). Most of their feeding time was spent scratching through humus beneath fig trees. Bantams released on an island off the Scottish coast showed an interesting age difference. Juvenile birds' food consisted mainly of invertebrates, presumably because, as growing birds, they required a high protein diet, while adults in the main ate cereal grains (oats) in the autumn and winter, grass and herbage in the spring and summer (Savory *et al.*, 1978). A wide survey of 21 galliform species showed that 16 of them consumed mostly animal food in the first two weeks of life; whereas by 8 weeks of age 20 out of 21 species were subsisting mainly on plant material – herbage or seeds (Savory, 1989b). Medium hybrid hens in small flocks on free range, with access to *ad libitum* mash inside their house, ate in addition a considerable quantity of grass from the pasture, estimated at 50 g dry matter/day (Hughes and Dun, 1983). In spite of its relatively high fibre content the grass would have made an appreciable contribution to the hens' energy requirements, while the carotene it contained coloured their egg yolks a deep yellow.

The feeding behaviour of other galliforms is very similar. By contrast, geese graze and ducks eat aquatic vegetation and invertebrates.

8.3   Development of feeding and drinking

Newly hatched chicks do not have an innate ability to recognize food, but they possess a strong propensity to peck at small particles, both nutritious and non-nutritious (Hogan, 1973). As time passes, however, pecking at inedible particles, such as sand, declines and pecking at food increases as they learn to respond to the consequences of consuming different items. Under natural conditions, chicks' attention is directed towards food because they follow their mother around and, whenever she stops to peck at a food item, they gather round and join in the pecking activity. The only practical system in which this would occur would be if chicks were naturally brooded in the case of a small farmyard flock. Feeding is, therefore, in some respects, a social activity and even when chicks are reared under commercial conditions they tend to feed as a group whenever possible. If they are isolated their food intake is depressed: chicks kept on their own eat less in the short term compared to chicks housed with a companion (Tolman and Wilson, 1965). This social facilitation has been put to use in stimulating feeding. Young turkey poults sometimes suffer from a condition called 'starve-out', during which they fail to start feeding. Putting in with them a broiler chick which is already feeding well can be very effective in encouraging the poults to eat (Savory, 1982).

As in the case of food, young chicks are initially unable to recognize water.

They have, however, a tendency to peck at flat, shiny surfaces. This results at them pecking at a pool of water and, as soon as their beak is immersed, they begin to learn to drink; the characteristic movement, during which the head is raised and swallowing occurs, is innate. Plain water is not, in fact, the most attractive stimulus for stimulating pecking. Other liquids, including blue-coloured water and, especially, mercury, are supernormal stimuli in this respect, at which they will peck in preference to normal water (Rheingold and Hess, 1957). Chicks under commercial conditions have some difficulty in learning to peck at nipple drinkers; this movement has to be learnt. For this reason the pressure in the system is often increased for the first few days, so that water drips slowly from the drinkers, thus encouraging chicks to peck at the shiny drops.

8.4 Physical characteristics of food

The factors which influence food intake include its physical characteristics such as particle size, colour, taste and smell and birds' familiarity with these. The effect of these features may be described as the food's palatability. It is not possible to quantify palatability except by measuring food intake, but some variations in preference cannot be accounted for by any other known factors (see section 8.5).

There is evidence that birds, given a choice, prefer particulate diets such as pellets to mash (Calet, 1965). The heating process which pellets have undergone may be a factor here, because reground pellets are also preferred to mash. Broilers given the same diet as either mash or in pelleted form ate more food when they were offered pellets and grew more quickly (Mastika and Cumming, 1981). However, food conversion efficiency is also better with pellets because more energy is spent eating mash (Savory, 1974). There is evidence that domestic chicks prefer a particle size of 2 to 3 mm (Bessei, 1973). In domestic hens, a tendency to select large particles (>2 mm) out of a mash diet has been reported (Perry *et al.*, 1976); these larger particles were of cereal origin and were high in energy and lower in protein. The preference for larger particles was especially clear in pullets; when they reached maturity they selected a greater proportion of small particles. This may reflect the laying hen's requirement for a higher protein diet as she comes into lay and needs the protein for albumen production. Not only do hens avoid the smallest particles in a diet, there is evidence that if a diet is too finely ground it can actually have harmful effects – the particles can accumulate as congealed masses in the oral cavity and pharynx, eventually causing infected lesions (Gentle, 1986b).

Poultry come to prefer the kind of food to which they are accustomed and a major change of diet can result in problems: unless the new diet is similar in texture, colour and probably taste to the previous one there may follow a reduction in food intake and therefore a check in growth rate or egg production. Under normal conditions the diet is changed at least twice as birds mature:

chicks of a laying strain, for example, are commonly switched from their starter diet at about 4 weeks to a grower diet and then the pullets are moved to a layers' diet at about 18 weeks. However, some companies in the USA use as many as seven different diets by 18 weeks.

In practice, diets are usually fed in the form of a mash. In the case of growing broilers, however, it is usual to feed chicks with crumbs or crumbles (friable, partly fragmented pellets), for about the first 3 weeks of life in order to give them a good start. The crumbs are initially presented in trays on the floor because the food is easier to find there, but after 4 days when the chicks are eating well food is provided in troughs. Growing broilers from about 3 weeks of age are generally fed on small pellets. If fed on mash, broilers take about 2 days longer to reach their final weight.

The resulting check from moving birds to a new diet has been utilized to induce a pause in production, for example in the UK, where complete deprivation of food is no longer permissible for this purpose (MAFF, 1987c). Instead, the hens are switched from a laying mash to a diet of whole grains, usually oats. The daily amount offered is limited to about 30 g and the birds initially find the new diet extremely unpalatable, do not consume even this amount and so rapidly cease laying (Lynn, 1989). The whole-grain diet is also deficient in protein and minerals and this is also an important factor in helping to terminate egg laying.

Other factors which affect food intake include its composition, social influences, energy requirements, lighting patterns and their associated diurnal rhythms. These are discussed in the following sections.

8.5   Diet selection

Under natural conditions wild birds are faced with an array of different food items which vary widely in nutritional composition; from these they are capable of selecting a diet which is adequate for all their requirements. The evidence is now strong (Hughes, 1984) that domestic birds, offered a range of different foodstuffs, can also choose a diet which provides them with all the nutrients necessary for growth, maintenance and production. To ingest an adequate diet they require a control system to allow them to select suitable amounts of each food. In domestic fowl and the other species of poultry this involves a 'generalist' approach (Rozin, 1976), in which they initially sample most potential foods and then continue to consume those which are palatable and nutritious. As discussed already Hogan (1973) has shown that newly hatched chicks at first peck equally freely at food or sand. By the time they are 3 days old, however, the ingestion of food leads to an increase in feeding behaviour within about an hour, whereas pecking at sand has no such effect. Thus, the food (but not the non-nutritious sand) reinforces the learning process.

Domestic fowls which are deficient in a particular nutrient such as calcium or sodium show an increase in generalized searching behaviour, pecking at objects which they would not normally investigate (Wood-Gush and Kare, 1966; Hughes and Whitehead, 1979). Food selection can be very precise; selective preference tests have shown that the fowl has specific appetites for such essential elements as calcium (Mongin and Saveur, 1979), phosphorus (Holcombe *et al.*, 1976a) and zinc (Hughes and Dewar, 1971), for vitamins such as thiamine (Hughes and Wood-Gush, 1971) and for protein (Holcombe *et al.*, 1976b). Rather surprisingly, there is no evidence that they preferentially select sodium-containing diets, even when deficient in the element (Hughes and Wood-Gush, 1971; Sykes, 1988), perhaps because under natural conditions such a deficiency is most unlikely, so the appropriate behaviour has not developed.

Domestic fowls thus have effective selection mechanisms and it has been argued (Emmans, 1975) that this ability to choose an appropriate diet can be exploited to increase dietary efficiency under commercial conditions. In large populations there is considerable variation in rates of egg production or growth. The more productive individuals require a diet higher in protein, minerals and vitamins than the less productive ones. If a complete diet is formulated to support maximum output, it provides expensive nutrients which are surplus to the requirements of the remainder of the flock and are therefore wasted (Figure 8.1). On the other hand, if a cheaper and somewhat less nutritious diet which meets the needs of the average bird is formulated, then the more productive birds will be unable to achieve their potential. This dilemma can be overcome by offering the diet in two (or more) portions, one suitable for maintenance and based on a cereal grain, so that it is low in cost and relatively high in energy, but low in protein, vitamins and minerals. The other complementary portion or

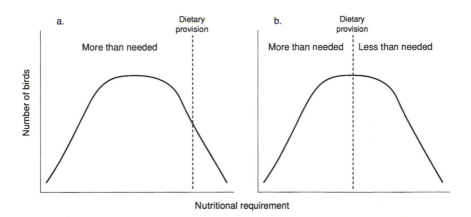

Figure 8.1. Diet and productivity. a. Diet formulated to meet the needs of the most productive birds will provide more nutrients than the majority of the flock require. b. Diet formulated to meet the mean requirements of a flock will be inadequate for the more highly productive bird.

'balancer', which is suitable for production, is expensive, but high in protein, vitamins and minerals. The approach is dependent on the proposition that, when offered two foods, birds will select a diet which allows them to produce as well as they would on the better food alone and will also avoid excess nutrient intake. The relative amounts of the two diets consumed by each individual bird should thus be a reflection of its potential output.

A number of experiments (reviewed by Hughes, 1984) have been carried out to test this proposition with growing chickens, growing turkeys and laying hens. The results are not entirely clear cut, but there is now convincing evidence that birds offered a choice between diets generally grow or lay as well as those given a single complete diet. However, the prediction from the theoretical model that there should be a gain in efficiency in flocks on a self-selection regime is not fully supported. Depending on the study, food utilization efficiency was inferior, similar or superior, for reasons which are not immediately clear.

Some of these experiments, such as those of Cowan *et al.* (1978), suggest that fowls do adjust their intake of protein and energy reasonably accurately as the nutrient content of diets is varied, but that palatability is also important. Wheat was preferred to barley even though nutritionally they were very similar, with only minor differences in metabolizable energy (ME) and fibre content, while the birds tended to overeat the balancer and thus consumed too much protein. It is known that laying hens prefer wheat to barley, oats or rye (Engelmann, 1940).

It is also important that the basal diet is more palatable than the balancer. Birds offered both diets will then select the basal portion preferentially but should consume sufficient of the balancer to meet their physiological requirements in terms of both maintenance and production.

Caged hens offered a choice in a split trough, with a whole grain/crushed pea mixture in the trough nearer to them and a protein balancer further away gave good results (Tauson *et al.*, 1991). They produced a greater egg mass but also ate more food, so the food conversion ratio was slightly poorer.

It is perhaps expecting too much to argue that modern hybrid strains should be able to select a precisely balanced diet from a number of components, in such a way as to maximize food conversion efficiency. These strains have now been maintained for many generations on a single adequate diet so may have lost some of their progenitors' ability to select their own diet. In any case there would have been little evolutionary pressure under natural conditions to select diets in such a way as to maintain maximum efficiency of their utilization. Other factors would have been important to birds faced with a wide range of foodstuffs, such as avoiding potentially toxic substances which often taste bitter or unpleasant; this may be why palatability appears to play an important role. Factors such as taste, colour, particle size and tactile characteristics of feedstuffs need to be given more attention. In order to take full advantage of self-selection regimes it might be necessary to select birds genetically for their ability to maintain high food utilization efficiency on self-choice diets.

8.6   Social influences

Social factors can have an important influence on feeding in adults as well as during development. Even in individual cages domestic hens tend to feed as a group, probably because the sight or sound of one bird feeding triggers feeding in others (Hughes, 1971). Under natural conditions it would be adaptive, because they would be attracted to join other feeding birds and thus increase their chances of finding food. This propensity towards group feeding does have implications for the provision of feeding space in intensive systems. Ideally, there should be room for all birds to feed at the same time, because at certain times of day a combination of diurnal rhythms and social effects is likely to mean that most birds are motivated to feed simultaneously. This is illustrated by the finding, as Figure 8.2 shows, that in wide, shallow cages where each hen

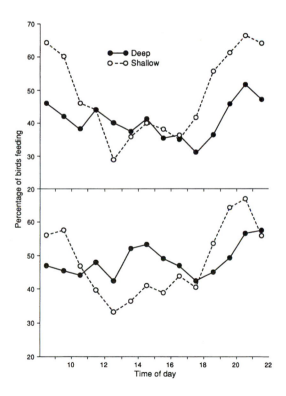

Figure 8.2.   Percentage of birds engaged in feeding activity at hourly intervals throughout the day in either wide, shallow cages, or conventionally narrow, deep cages. The upper graph shows the pattern when there were four birds per cage, the lower graph when there were three per cage. In the shallow cages more birds fed at the peak times of morning and evening (Hughes and Black, 1976).

had 15 cm of feeding space, birds tended to feed as a group (Hughes and Black, 1976). In contrast, in conventional cages with 10 cm of space fewer birds fed at a time but feeding activity extended over a greater proportion of the day; those birds which couldn't feed at peak times had to wait until overall activity had fallen.

When the food troughs of groups of three hens in floor pens were partitioned into three separate feeding areas they spent less time feeding and ate less food than controls with undivided troughs (Huon *et al*., 1986). In a similar manner, subdividing the feeding space of caged hens by placing dividers along the food trough, thus partitioning it into smaller segments, reduced the time they spent feeding and also the number of agonistic interactions at the food (Preston and Mulder, 1989). This suggests that providing enclosed feeding space may be advantageous in allowing hens to feed with less disturbance and influence from other birds than is the case with the normal open troughs. The only possible disadvantage is that it also appears to reduce time spent manipulating the food (section 8.9). This may have implications in that it releases more 'free time' which, if directed towards other birds, may increase the amount of feather pecking.

8.7   Temporal patterning of feeding

Feeding behaviour does not occur at random. It is organized in the short term into bouts or meals and, on a longer term basis, generally shows a clear diurnal rhythm, being unevenly spread throughout the day. This topic has been well reviewed by Savory (1979).

The sizes of meals and the intervals between them when no feeding occurs have been studied both in domestic fowl and in quails (Figure 8.3). There are clear relationships: large meals tend to be preceded by long intervals and, even more so, followed by long intervals. For small meals, the preceding and following intervals tend to be brief. This shows that meal-eating is governed by both hunger and satiety mechanisms (Savory, 1979), with signals from crop, gizzard or duodenum playing an important part.

In young chicks, which are often kept on very long photoperiods such as 23 h light : 1 h dark, there is little or no periodicity of feeding in the first week of

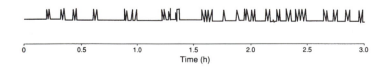

Figure 8.3.   Feeding activity of Japanese quail (after Savory, 1980) organized into separate bouts, separated by intervals when no feeding occurs.

life, even though they hatch with an inherent circadian rhythm of about 25 to
25.5 h (Aschoff and Meyer-Lohmann, 1954) but they then gradually develop a
diurnal rhythm, especially if moved on to shorter photoperiods. Laying birds
kept on a 14 to 17 h photoperiod usually show very marked rhythms, with one
feeding peak in the morning and a more pronounced one towards the end of the
day. Birds do not feed in the dark when the photoperiod is adequate (greater
than about 6 to 8 h), so the morning peak is presumably caused by refilling of
the crop, which acts as a food reservoir and will have emptied overnight. In the
evening birds fill their crops in order to have enough food to last until the next
morning. The evening peak is usually less in non-laying birds, but it is greater
when a signal of impending darkness, such as a period of artificial dusk, is
given (Savory, 1976), as Figure 8.4 shows. This suggests that the birds are
inclined to feed at this time but, in the absence of a suitable cue, have difficulty
in anticipating the onset of darkness. The peak may be more obvious in laying
birds because they receive internal physiological cues from the process of egg
formation. On a 14 h photoperiod the timing of the evening peak correlates
roughly with the start of shell calcification and an increase in calcium requirements, which results in an increase in food intake at the end of the day, but only

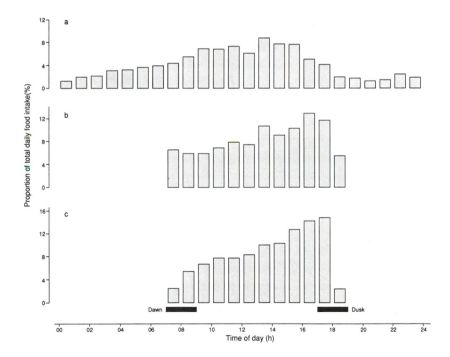

Figure 8.4. Diurnal patterns of food intake of young cockerels showing how they
are influenced by lighting pattern (after Savory, 1976). a. Continuous lighting. b.
12 h of uniform light intensity. c. 12 h of light with a 'dawn' and a 'dusk'.

on days when an egg is being formed (Hughes, 1972; Mongin and Saveur, 1974).

A number of other factors affect feeding patterns. Laying hens and quail show reduced food intake for about two hours before oviposition, followed by a compensatory increase afterwards (Woodard and Wilson, 1970). The reduction is not simply due to other behaviour patterns, such as nest investigation and sitting, keeping them away from the food trough. Even in cages, where the trough is nearby and the hen may peck at food from time to time, intake decreases similarly, suggesting a specific reduction in feeding motivation during the pre-laying period.

The form and density of the diet have an influence. Although total food intake was similar in both cases, hens given pellets rather than mash displayed a more pronounced diurnal rhythm (Fujita, 1973). This was because they took longer to consume a given weight of food when it was presented as mash rather than pellets, so feeding time occupied a larger proportion of the day, thus tending to blur the underlying pattern (Figure 8.5). In the same way, reducing the nutrient density of a diet, for example by diluting it with an indigestible filler such as cellulose powder, also increases total feeding time and again minimizes the diurnal pattern (Savory, 1980). In this case, of course, the weight of food consumed increases to compensate for the dilution. Increasing the time spent feeding by altering the nature of the diet can, in some circumstances, be an

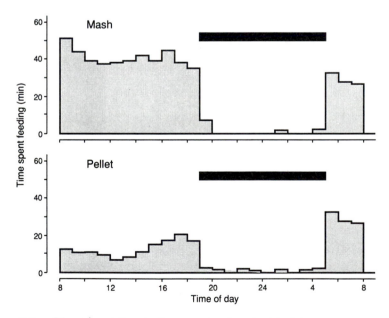

Figure 8.5.   Diurnal variation in time spent eating; the black bar represents darkness. Hens receiving mash spent longer feeding and showed a less pronounced diurnal rhythm than birds receiving pellets (after Fujita, 1973).

advantage, for example by reducing the danger of feather pecking and cannibalism (sections 9.8 and 9.9). This effect is presumably achieved because of the extent to which pecking activity is directed towards food rather than the plumage of other birds.

Birds do not normally feed during the dark period but will do so if the photoperiod is very short, for instance 6 h or less (Morris, 1967). Intermittent lighting patterns are gaining increasing acceptance for commercial reasons (Morris *et al.*, 1988) and birds respond by modifying their feeding activity appropriately. Intermittently lit hens housed in cages performed 25% less feeding activity than hens on a 16 h photoperiod but consumed similar amounts of food. They fed during the 45 min dark periods alternating with 15 min light periods, but did not feed during the longer 8 h dark period, presumably interpreting this as night time (Lewis *et al.*, 1987). In non-cage systems, however, it is less likely that birds would feed during the dark periods because they might have problems locating the food troughs.

8.8   Food intake

Because production is generally similar across a broad range of housing systems the amount of food eaten in different systems depends on three main factors – wastage, which is determined primarily by food trough design; energy requirements, which are influenced by ambient temperature, activity of the birds, feather covering and body weight; and nutrient density of the diet provided. Comparative values, all relating to ISA Brown medium hybrids, suggest that mean daily intake can be around 118 to 120 g/day in cages (and often lower without affecting performance) and considerably more in non-cage systems (Figure 8.6): 130 g/day on deep litter (Appleby *et al.*, 1988b), 143 g/day in a covered strawyard (Gibson *et al.*, 1988) and around 140 g/day, supplemented with about 50 g of grass from the pasture, on free range (Hughes and Dun, 1983).

Males and females of the same strain have different food requirements associated with their different body sizes. With broiler breeders, however, males and females can be reared together by fixing a grid over the main feeding trough with gaps wide enough to allow females to feed but not males, which have wider heads. Troughs for the males are then hung at a height which only they can reach. Care must be taken in selecting grids for the head size of the appropriate strain, because if the gaps are too narrow some hens suffer facial lesions and swelling (Duff *et al.*, 1989).

8.9   Foraging behaviour

There is increasing evidence that foraging behaviour is important to poultry. Under semi-natural conditions jungle fowl, even though they were fed regularly,

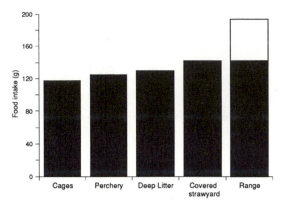

Figure 8.6. Daily food intake (layers' mash) of the same strain of medium-hybrid laying hens kept in five different housing system. The clear portion in the case of range is an estimate of their supplementary intake of herbage from the pasture.

allocated a large proportion of their time to foraging activities (Dawkins, 1989). In non-cage systems, such as covered strawyards or deep litter, foraging in the form of scratching or pecking at material on the ground occupied between 7 and 25% of birds' time (Gibson *et al.*, 1988; Appleby *et al.*, 1989). In cages birds have no access to loose material such as litter but instead spend a substantial proportion of their time either feeding or manipulating the food in the trough with their beaks. The manipulation takes two main forms: food is either drawn towards the birds and piled up at the back of the trough or is flicked back and forth with vigorous beak movements, some of it ending up outside the trough and being wasted (Figure 8.7). These movements probably represent the appetitive component of foraging behaviour, which the birds carry out in the food because it is the only substrate to which they have access.

Food wastage through manipulation can be economically important so a number of commercial techniques have been adopted to minimize it (Elson, 1979). These include a wire grid at the level of the food, so birds have to peck through the spaces in order to feed, a spiral along the bottom of the trough which prevents flicking, or relatively deep, narrow troughs with a shallow depth of food replenished by an automatic chain or other conveyor running in the base of the troughs (Figure 8.8). If the ability to perform appetitive behaviour is important then it is possible that wastage-reducing methods could be a source of frustration to caged hens. To safeguard welfare it may be necessary to provide them with a suitable substrate such as the sand bath now being added to some novel designs of modified cages (section 4.10). Alternative systems to laying cages mostly include an area for foraging, usually of wood shavings, but sometimes of fine gravel, as in the Elson Terrace. In these cases no conflict between the need to minimize food wastage and the need to provide a source of activity for the birds arises, because other outlets for foraging behaviour are available.

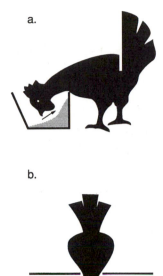

Figure 8.7. Two types of food manipulation performed by caged hens, which can result in food being pulled or flicked out of the trough, resulting in wastage. a. Beak drawn back towards body. b. Beak flicked from side to side, scattering food.

8.10   Drinking behaviour

Adult domestic fowls drink about 150–200 ml of water per day at normal ambient temperatures. This quantity can be consumed in a relatively short time: Gibson *et al.* (1988) found that hens in a covered strawyard spent about 6% of the photoperiod in drinking behaviour. However, this proportion of time can be much longer in cages. Bessei (1986) reported that caged hens spent, on average, 8 min of each hour (14%) engaged in drinking behaviour. Drinking is generally closely associated with feeding (Hill *et al.*, 1979) and there is often a clear circadian pattern (Wood-Gush, 1959b), with an increase in water consumption towards the end of the day because of the evening feeding peak.

For laying hens in floor systems one circular drinking fountain per 100 birds is recommended. This provides about 1.2 cm/bird of drinking space. Alternatively, one nipple drinker with drip cup may be provided for every 10 birds. In battery cages containing up to six birds, one nipple drinker is generally regarded as sufficient, but the UK Codes of Recommendations (MAFF, 1987a) recommend that every bird shall have access to two nipples in case one should become ineffective, and this is achieved by placing them at the boundary between two cages. Although the relatively brief time spent at the drinker would imply that competition for space is unlikely, in fact there is evidence that both the amount of drinking space and the way in which water is presented can influence intake. As the number of birds per nipple drinker is reduced their water intake goes up. Hearn (1976) found that daily water intake per bird was 165 ml when the number of birds/nipple was 10, 169 ml when there were 5 and 182 ml when there were 2.5 (Figure 8.9). Intake increased still further to 213 ml per day when water was supplied in troughs. However, this modest constraint had no effect upon egg output.

Because drinking from nipples is not a natural behaviour, birds develop a number of different strategies for obtaining water from them (Hill, 1977): some peck at the nipple, some hold the plunger up and drink the water as it flows over their beak, some peck at water droplets on the cage structure and some prefer to drink from the drip cups under the nipples. Sometimes birds are reared from day-old in cages with nipple drinkers and then, at point-of-lay, are moved to a floor system with trough drinkers. At this age, unlike chicks, they no longer possess the tendency to peck at flat, shiny surfaces; there is thus a danger that

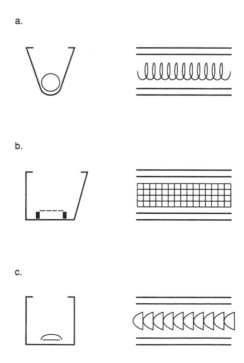

Figure 8.8.   Food trough designs which reduce wastage. a. Fixed spiral. b. Fixed grid. c. Moving chain which both distributes food and minimizes wastage.

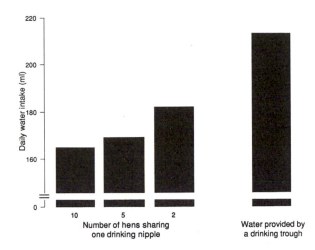

Figure 8.9.   Water intake of hens increases as the number of birds per nipple drinker gets less; it is higher still when water is supplied in a trough.

they may not recognize the water and so fail to drink. Even in chicks, dipping the beak into water results in them starting to drink earlier, presumably because they learn to recognize it more quickly (Yeomans, 1987).

In some litter-based systems, particularly in the case of broilers, problems can arise with wet litter. This may be due to inadequate ventilation and poorly designed drinkers where, at very high stocking densities, pushing and jostling beween birds causes the drinkers to be tipped and water to spill. The wet litter can result in hockburn, necrosis of the feet and breast blisters (Figure 8.10). Improved drinker designs help here, together with increased ventilation to evaporate spilled water as well as that excreted by the birds; it may also be advisable to reduce the stocking density. Other possible causes are nutritional: a diet which is too high in minerals, especially sodium or potassium, can lead to overdrinking, while high-fat diets have also been implicated.

8.11   Drinking as a stress-related behaviour

There is growing evidence that birds drink not only to meet their physiological requirements but also as a response to stress. Lintern-Moore (1972) observed that a certain proportion of caged laying hens produced particularly wet droppings and found that the water consumption of individuals with wet droppings was almost three times as much as that of normal hens. She concluded that the cause was behavioural polydipsia or psychogenic overdrinking and speculated that 'boredom might cause some birds to drink excessively as a sort of conditioned response to their captivity'. Dosing caged hens with tranquillizing or opioid-blocking drugs reduces water intake without affecting food intake,

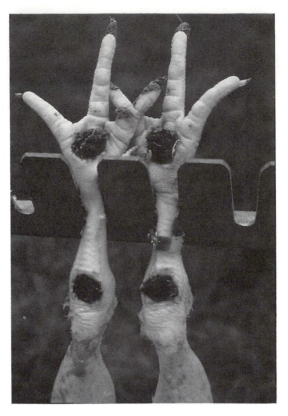

Figure 8.10.   Lesions such as hockburn and foot necrosis can be associated with the presence of wet litter.

suggesting there may be an element of overdrinking by most fowls (Savory *et al.*, 1989). The repertoire of agitated behaviour which precedes laying in some caged hens often includes a drinker-oriented element, during which hens typically turn towards the drinker, peck at it or touch it with their beaks, turn away and carry out some other behaviour, then turn back and go through the same repetitive routine again and again.

Birds which are food-deprived may also show an increased incidence of drinking behaviour (Savory *et al.*, 1992). Broiler breeder hens have to be kept on a restricted feeding schedule – when given food *ad libitum* they overeat, become obese and show reduced fertility because of hyper-ovulation (Hocking *et al.*, 1989). If these food-restricted hens are supplied with water *ad libitum* they may drink excessively large quantities; this results in wet droppings, wet litter, high concentrations of atmospheric ammonia and unpleasant conditions both for birds, causing foot, hock and breast lesions, and for human operators.

To avoid these consequences the breeders' recommendations often suggest that the water supply is limited to only a few hours a day to prevent overdrinking and to reduce the opportunities for spillage. This may, however, exacerbate the frustration caused by food restriction.

Food and water are supplied to poultry in equipment which has been designed primarily from the viewpoint of keeping the food and water clean and fresh with minimal wastage. This is in the interests of both the producer and the bird, because it helps to avoid contamination and possible disease and reduces environmental problems attributable to wet litter. The only conflict that might arise is concerned with food wastage prevention, which, in the barren environment of the battery cage, may in some circumstances frustrate foraging behaviour sufficiently to have welfare implications.

9

Social behaviour

9.1 Summary

- Social behaviour is behaviour in relation to other animals; this chapter covers all such aspects of behaviour other than reproduction. Fowls under natural conditions are found only in small groups, whereas ducks and geese sometimes form very large aggregations.
- Chicks, ducklings and goslings learn the characteristics of conspecifics through imprinting and other behaviour patterns such as foraging by imitation of maternal behaviour. Under commercial conditions, with no model to follow, they sometimes fail to learn important behaviour patterns such as perching; failing to eat or drink is a problem in turkey poults.
- Synchrony is common for a number of behaviour patterns, including feeding, preening and resting; systems should be designed to take this into consideration.
- In the smallish groups typical of cages or small pens, dominance hierarchies form; overt aggression is unusual except at times of disturbance. In most alternative systems with large groups, individuals move extensively through the house rather than forming smaller sub-groups, these constant encounters with strangers may be a source of stress. Especially submissive individuals which are severely pecked have been identified; they represent a major welfare problem.
- Feather pecking occurs in all systems but is usually worse in cages. Cannibalism causes severe suffering, is commoner in non-cage systems and is often unpredictable in occurrence. Both are generally ameliorated by reducing light intensities, by providing opportunities for foraging and by beak trimming. However, beak trimming is itself undesirable for welfare reasons.
- Occurrence of harmful social behaviour is strongly influenced by the

environment and can therefore be reduced by management. Use of the term 'vice' for such behaviour is inappropriate.

9.2 Natural behaviour

In the wild, jungle fowl live in small groups. The most common groupings seen are of several females with one male, with other males being solitary or in small groups (Collias and Collias, 1967). Each group has a regular roost and an area in which it usually forages. Other galliforms are similar in their social behaviour and the same situation is found in feral domestic fowl, which form distinct social groups, each with a home range (Wood-Gush *et al.*, 1978). Conditions are also similar for small farmyard flocks and social behaviour in these is probably very like that of wild birds.

Wild mallard ducks and greylag geese are partially or wholly migratory and form less stable social groups than the galliforms. They frequently occur in large aggregations, so in this sense they are better adapted to domestication. However, geese in particular show individual recognition at least to the extent of long-term sexual pairings, so this is one aspect of behaviour which is disrupted in domestic conditions.

9.3 Socialization

In farmyard flocks, young poultry usually grow up in a group of mixed age and sex. They are mobile from hatching, so it is important that they learn to recognize their mother and their siblings and this occurs through a process called imprinting. After hatching, they instinctively follow the first moving objects they see and learn their characteristics. As the mother broods them and helps them to find food they subsequently learn the advantages of staying with her and with the rest of the brood. Maternal imprinting can occur only during a sensitive period of about two days after hatching, so in normal domestic conditions where the mother is absent, this learning process is restricted to learning the features of hatch-mates. Sexual imprinting, in which birds learn the characteristics of potential mates, follows at a later age. It is more gradual and in chickens the ages of about 10 to 12 weeks are most important (Siegel and Siegel, 1964). When the sexes are reared separately, subsequent mating is likely to be less efficient.

Behaviour learned from the mother includes foraging and use of the home range area, but birds can learn to feed on their own (section 8.3). In galliforms, perching behaviour is also influenced by the mother. In chickens, the mother broods the chicks on the ground until they are 7 or 8 weeks old, then resumes roosting in trees or bushes, initially low down and later with other adults. The chicks jump up to follow the mother (Wood-Gush *et al.*, 1978) and can subsequently perch on their own. In the single-age groups of commercial brood-ers and rearing pens, maternal teaching of behaviour is not possible and surpris-ingly often birds fail to learn appropriate behaviour from each other. Thus, if medium weight chickens are reared without perches, only some individuals will learn to perch as adults, while others in the same groups fail to do so (Figure 9.1). The latter birds have difficulty reaching raised drinkers or nest boxes (Appleby *et al.*, 1983). Similarly, turkey poults may fail to drink and may die of dehydration despite flock-mates drinking nearby. These problems can usually be alleviated, however, once recognized. In the case of perching, provision of

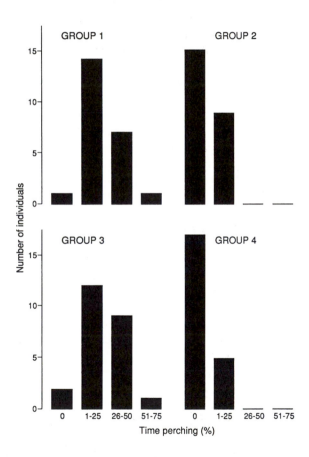

Figure 9.1. Effect of early experience on perching by adult hens. In this experi-ment, perching by individual medium hybrids was recorded shortly after they were moved at 20-weeks-old to pens with raised nest boxes. Groups 1 and 3 had been reared from 5 weeks with perches, groups 2 and 4 without. Many of groups 2 and 4 failed to reach the nests and laid on the floor (Appleby *et al.*, 1983).

perches during early rearing usually results in all chicks learning to perch (Appleby *et al.*, 1983, 1988a). To encourage drinking by young turkeys, some chicks are often included in the flock, as the poults can apparently learn drinking from these even if not from each other.

9.4 Behavioural synchrony

Although birds do not always learn behaviour from each other, they readily copy behaviour in such a way as to perform different activities synchronously in a group. This is particularly clear with feeding and in all husbandry systems it is common for birds to feed together rather than at independent times. Some of this synchrony can be accounted for by factors which act separately on all birds, for example the light regime, which produces peaks of feeding at dawn and dusk quite apart from social influences. In addition, though, bouts of feeding through the day are more synchronized between birds than would be expected at random (Hughes, 1971). This pattern is constrained for hens in cages which have the minimum requirement of 10 cm food trough per bird, not so much because of the feeding space itself as because the cage is too narrow for all hens to stand side by side. This results in hens contending to feed simultaneously. The problem is avoided in shallow, wider cages (Figure 9.2). It is also less in large flocks fed *ad libitum*, where synchrony is unlikely to be complete. It is worse in breeding flocks on restricted food, in which all individuals react to the arrival of a new delivery of food. Round feeders allow more birds to feed simultaneously than linear troughs, for the same actual feeding space.

Synchrony in resting is obviously also influenced by the light regime, but also occurs at periods in the day, interspersed with bouts of feeding. It is advantageous in cold conditions, when roosting in contact conserves body heat. Again, roosting side by side is constrained for hens in narrow cages, but no problem in wide cages or floor systems. At night, in such systems as deep litter and aviaries, 'rafts' of hens form in close body contact. This happens even when lights go off abruptly, because films taken by infrared light have shown birds shuffling together, forming groups by touch. However, roosting behaviour is facilitated in houses with a gradual dusk, or where lights-off is preceded by a dim light period. This is particularly important when roosting above ground level, on perches or on a platform, is to be encouraged. In houses with partially slatted floors, this will be appropriate for the accumulation of droppings produced during the night.

Other behaviour patterns, such as drinking and preening, also tend to be performed synchronously, perhaps partly because they occur between feeding and resting bouts. As with feeding, synchrony is greater in the small groups of laying hens housed in cages, but may also be restricted in cages. Preening requires considerably more space than the EC minimum requirement of 450 cm^2 per bird (Table 12.1), so not all birds in a cage can preen simultaneously.

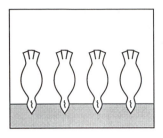

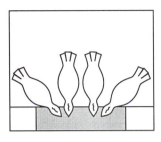

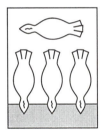

Figure 9.2.   Examples of the orientation of feeding birds when given adequate cage and trough width (top), adequate cage width but restricted trough width (middle) and restricted cage and trough width (bottom) (Hughes, 1983a).

9.5   Affiliative behaviour

The main affiliative behaviour shown by poultry is flocking. The tendency to form groups rather than move independently or avoid other members of the species evolved primarily for protection against predators. Even in the absence of predators, birds in large areas generally clump together. This is easily observable in floor housing systems and particularly clear in free range. Birds which go out onto pasture from a free range house generally move as a flock. They often stay near the house (Figure 9.3), which may be partly due to attraction to the rest of the flock in the house.

A behaviour pattern of more immediate mutual advantage is the habit of pecking food which has adhered to the face of another bird. The bird being pecked remains very still, often with its head back and its eyes closed, allowing the pecking to continue. This probably happens more often at high stocking density where birds are feeding in close proximity. In hens, it may also happen more often in cages than in other systems, because of the lack of suitable objects on which birds could clean their own faces, by beak-wiping. There is a possibility that this could produce a disadvantage despite the obvious advantage. Such behaviour may be a pre-disposing factor to feather pecking or cannibalism, because birds being pecked in those cases frequently also freeze, rather than trying to escape.

9.6   Group size and spacing behaviour

Group size has effects independent from stocking density. First, at the same density a larger group will have a larger absolute space in which to move: hence

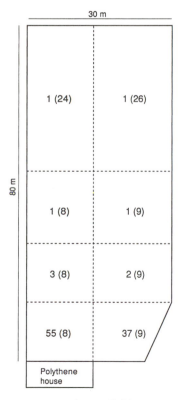

Figure 9.3.   Birds in free range use the available pasture unevenly, usually tending to stay near the house. This diagram shows the percentage use of different parts of a pasture in one study; figures in parentheses show percentage of total area (Keeling *et al.*, 1988).

the UK regulations for small groups of hens in cages (section 4.9). Second, in large groups there are more individuals with which to interact and in all captive animals frequency of social interactions increases with group size. Third, birds in large groups will have more difficulty learning to recognize their flock-mates individually. Individual recognition in hens seems to be limited to groups of up to about 80 birds (Guhl, 1953): this range includes most farmyard flocks and get-away cages and is similar to the flock size of small units in percheries or slatted floor systems. It is exceeded in most floor housing. In general, small group size is advantageous. For example, in cages for laying hens small groups show higher production levels compared to larger unit sizes (Hughes, 1975b). There is also evidence that in cages, stress increases linearly with group size (Mashaly *et al.*, 1984; Roush et al., 1984). Modified cages which retain similar, small group size may have similar advantages (Robertson *et al.*, 1989). Some of these effects may be due to the behavioural problems considered in the following sections.

In large pens or houses, birds do not use the area evenly (section 12.5), but they generally move about sufficiently to suggest that they encounter all other members of the flock. This has been found for laying hens in strawyards (Gibson *et al.*, 1988) and on deep litter (Appleby *et al.*, 1989) and for broiler breeders on deep litter (Figure 9.4). Individual recognition is not possible in these conditions. It is not known whether birds become used to these continual encounters with unfamiliar individuals, but in small groups contact with strangers results in increased heart rate (Candland *et al.*, 1969), increased aggression (Craig *et al.*, 1969) and growth of the adrenal glands (Siegel and Siegel, 1961). In some houses, though, sub-groups may form within pens. For example, in one experimental aviary for laying hens there were nest boxes down the middle of the house (Hill, 1983) and although birds were not prevented from moving over these they rarely did so. Other designs of partial barriers in pens would also encourage the formation of sub-groups and research on such methods would be worthwhile. One recent suggestion is to imprint chicks on different colours or objects and then spread such cues round the house (Nicol and Dawkins, 1990).

Within groups, the spacing of birds varies with the activity that they are performing. While they will roost in body contact and preen quite close together, they are usually more spread out (subject to the constraints of their housing) while foraging for food on pasture or in litter. This is mostly a mechanical effect, because pecking and scratching result in a certain separation. However, it may also be partly a social effect, as sometimes one bird approaching another provokes aggression or retreat. This has sometimes been described as defence of a 'personal space' by each bird, but this is not as clear-cut as in some other animals: for example, any such tendency is clearly absent during roosting. Nevertheless, while birds do clump into flocks, as described in the previous section, they react to high stocking density by spacing more evenly than random. Thus in cages for laying hens, where close proximity is enforced, birds attempt

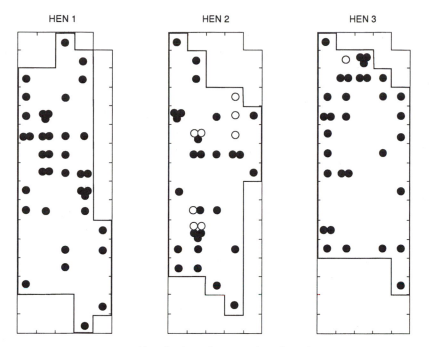

HEN 1   HEN 2   HEN 3

Figure 9.4. Movement of broiler breeders in a deep litter house, 46 × 15 m. Birds were chosen at random from a flock of nearly 4000, and their locations recorded on 52 days over 34 weeks; their ranges were all more than half the area of the house. ○ = nesting, ● = other records (Appleby *et al.*, 1985).

to stay further apart than random. This even occurred in experimental pens for three hens which provided 1400 cm^2 each (Keeling and Duncan, 1989). It is not known how this sort of spacing is affected by use of three dimensions in housing. For example, in a perchery or aviary hens are in close proximity above and below each other as well as horizontally, and it is possible that this also causes some stress. While vertical movement may be common in the wild, as when jungle fowl or pheasants roost in trees or bushes, this would not be at densities comparable to those in commercial systems.

9.7 Aggression and dominance

In a farmyard flock, as in a wild group, the first aggression experienced by young birds is probably that received from other members of the flock, when they move too close, or when they are in the way of older birds. Later, particular chicks, ducklings or poults themselves become aggressive to contemporaries. There are also usually certain individuals, perhaps smaller or weaker than others, which are attacked particularly frequently. However, after subsequent hatches there are younger birds in the flock which they attack in turn. Birds quickly learn that they should avoid others which will obviously beat them, but that they will be able to beat others which are much smaller or weaker. In these circumstances, the most common form of aggression is pecks to the head of the opponent. As the recipient tries to escape, it often receives these pecks on the back of the head. Birds which are more evenly matched are more likely to fight, in face-to-face encounters. However, if the group is small enough for members to recognize each other individually, they remember the results of such fights and avoid fighting with others which have beaten them previously. A relationship between two individuals in which one (the subordinate) avoids confrontation with the other (the dominant) is called social dominance and the set of such relationships in a group is called a dominance hierarchy or peck order. Social dominance was, in fact, described first in chickens, by a Norwegian called Thorlief Schjelderup-Ebbe working with small groups of hens in 1922. Some individuals high in the hierarchy are able to peck or displace many others, while some individuals low in the hierarchy are frequently displaced (Figure 9.5). In a small, stable group, however, actual aggression is usually rare, unless it is provoked by special circumstances such as restricted feeding space, because subordinates avoid dominants whenever possible.

In the groups of birds of the same age which are universal in commercial conditions, there is no equivalent movement up through a hierarchy with time. As described above, some birds become particularly aggressive and some receive more aggression. The effects of this, though, depend on group size and stocking density. If individual recognition is possible, a hierarchy will form: this occurs for laying hens in cages, wire-floored Pennsylvania systems or small perchery pens and for ducks or turkeys in small, penned groups. In larger groups, including most other housing systems for hens, there is no proper hierarchy. Some birds continue to be aggressive and some to be submissive, while the rest avoid the aggressive ones and join in the attacks on their submissive flock-mates. The difference between these types is reflected in their posture and movement, indicating confidence or the lack of it (Figure 9.6; Wood-Gush, 1971).

Among laying hens, the frequency of aggression is generally low in conventional cages, for at least two reasons. First, a group of four or five birds knows each other well and either has clear dominance relationships or accepts equal status. Second, if there is one clearly dominant bird this tends to suppress interactions between the others (Hughes and Wood-Gush, 1977). Aggression is more frequent, though, during times of disturbance such as pre-laying activity. The incidence of aggressive behaviour is higher in most alternative systems. It tends to decline as stocking density increases, perhaps because bird movements become restricted and because, as in cages, subordinates have to remain in close proximity to dominants. The perchery system with litter may be unusual in that the number of aggressive interactions per bird recorded was lower than in a

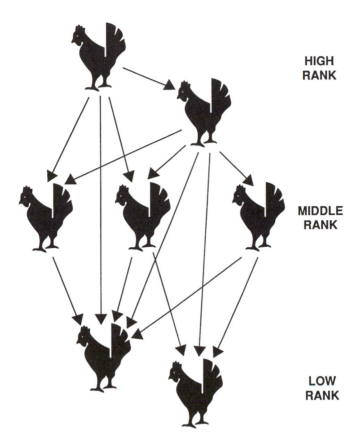

HIGH
RANK

MIDDLE
RANK

LOW
RANK

Figure 9.5.   A dominance hierarchy, or peck order. Birds of high rank are able to peck or displace many others (shown by arrows) while those of low rank are displaced by most or all of their flockmates.

Figure 9.6.   Posture and movement of birds show the difference between individuals which are confident or aggressive and those which are not.

comparison flock housed in cages (McLean *et al.*, 1986), perhaps because hens were able to withdraw from potential interactions in three dimensions. On the other hand, this was also the rationale behind the design and naming of get-away cages, yet ironically these usually have problems with some birds being bullied by others.

It is common in any large group of hens kept moderately intensively that a small number of birds, those mentioned above as being particularly submissive, will be pecked continually by others (McBride, 1958; Gibson *et al.*, 1988). This has been described as the 'peck order effect' (Duncan, 1978) and such birds as 'runts' (Appleby, 1985) or 'pariahs'. They have heads and combs scarred from pecking, poor body condition and posture and they spend most of the time trying to avoid interaction with others. This often means that they feed very little and they usually stop laying. As such, it is to the advantage of both the birds concerned and the producer if they are identified and removed. When isolated, they will resume feeding and laying as normal. The effect is less common in conventional cages or equivalent small groups in modified cages. It is probably more frequent in larger caged groups, because average individual production declines with group size in conventional cages (Hughes, 1975b).

If aggression is frequent overall, it can be reduced, along with other activity, by dim lighting. This option is only available in fully enclosed houses; not, for example, in strawyards with partially open sides. The other main management technique which reduces the effects of aggressive pecking is beak trimming, discussed in section 9.9.

Aggressive pecks sometimes break the skin on the head or comb. Injured birds should be isolated quickly, both to avoid further injury and because pecking in such circumstances can lead to cannibalism. If they cannot be isolated and the injury is insufficient to justify culling, daubing tar or a proprietary equivalent around the wound is sometimes effective in inhibiting further pecking by other birds.

9.8   Feather pecking

The problem of feather loss in chickens, and to a lesser extent in turkeys and pheasants, has received more attention in consideration of different housing systems than almost any other issue. The importance of feathers to the welfare of birds is not immediately obvious, so the main reason for this attention was probably initially aesthetic: birds with extensive feather loss are unattractive. However, such loss (apart from moulting) also indicates major behavioural or physiological departures from natural conditions and increases the danger of exposed skin being injured. The main cause of feather loss in all systems is not physiological change or abrasion but feather pecking (Hughes, 1985) and it is painful for a bird to have a feather pulled out (section 1.18). Feather loss can also be an economic problem: birds with few feathers lose heat faster and thus cost more to feed.

Feather pecking is different from aggressive pecking, both in character and in effect. The movements involved are not rapid and violent, as in aggression, but deliberate. They are similar in type to feeding movements (Wennrich, 1975), taking hold of a feather and then pulling, and in fact this behaviour is sometimes called feather picking. Feathers which have been removed are then sometimes eaten. Pecking is often directed at feathers which are marked or distinctive, or which are out of line. The most common area to be pecked, at least initially, is the back, perhaps because feathers out of line in more accessible places are quickly preened. Damage can then progress to the tail and even to the whole body (Figure 9.7). Feather loss from aggressive pecking is usually confined to the head.

Figure 9.7. An extreme case of feather loss. While some feather loss is caused by abrasion, most is by feather pecking.

Among hens, some individuals are particularly liable to feather peck others while some are prone to being pecked (Cuthbertson, 1980). There may be others in neither category, but they readily copy the behaviour of feather pecking and it can spread rapidly in a group. There are also differences between strains in incidence of feather pecking (Hughes and Duncan, 1972) and the implication of a genetic influence on the behaviour is supported by analysis of heritability in individuals (Cuthbertson, 1980). There have been experimental studies of selection against feather pecking in quail but more research is needed to show whether heritability in poultry is strong enough to make this practical commercially. There are also major environmental influences on the behaviour. It is worse in barren conditions, presumably because the availability of other, varied stimuli for pecking is then reduced (Blokhuis, 1989). It is therefore often a major problem in cages, reflected in the fact that worse feather loss has often been recorded in cages than in other systems (McLean *et al.*, 1986; Appleby *et al.*, 1988b). For example, loss in a perchery was 2.7 compared to 4.4 in cages (on a scale from 0 to 20; McLean *et al.*, 1986). In a study of free range, feather damage was also less than in cages. Scores over 4 years were 1.2 to 1.5 compared to 1.8 to 3.5 in cages (on a scale of 1 to 5; Hughes and Dun, 1986). However, severe feather loss may also occur in non-cage systems on occasion. This may sometimes be explained by an arrangement which allows some birds to defecate on others. In these circumstances, hens peck at soiled feathers and pecking may then spread. In one aviary system which had such an arrangement of tiers, very severe loss was recorded in the first few flocks housed (Hill, 1983) and similar problems have been recorded in get-away cages. However, outbreaks of feather pecking are not confined to systems where this is a possibility, since they have also occurred in Pennsylvania systems and, recently, in the Elson Tiered Terrace (Elson, 1989). High stocking density may be a contributing factor in such cases (Figure 9.8).

As already pointed out, feather pecking is similar to food pecking (Wennrich, 1975) and it is exacerbated by some feeding methods. Poultry can obtain sufficient food much more quickly in domestic conditions than in feral condi-

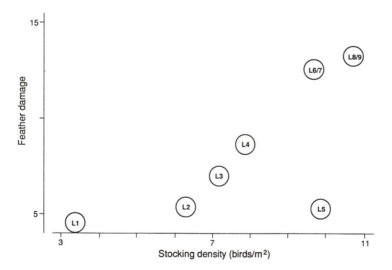

Figure 9.8.   The effect of stocking density on feather damage in a deep litter house for laying hens. Damage was scored at the end of lay on a scale from 0 (no damage) to 20 (denuded). Points represent different flocks, and despite the low damage in flock 5 the effect was statistically significant (Appleby *et al.*, 1988b).

tions, where birds spend 50% or more of their time in feeding (Savory *et al.*, 1978). Some of the remaining time may be spent in feather pecking. However, where pasture, litter or other loose material is available birds will forage even if this yields little food (Hughes and Dun, 1986; Gibson *et al.*, 1988) and this probably contributes to the difference between cages and other systems. Another major feature of feeding regimes is whether pellets or mash are supplied. This used to be determined by the method of automatic food distribution in a particular system. However, food in the form of pellets can be eaten faster than mash and this also encouraged feather pecking. For this reason, the use of pellets for laying hens is now rare.

As with aggression and all other activity, feather pecking can be reduced in enclosed houses by lower light intensity. However, light is usually kept at the lowest level compatible with production in any case, so this is available only as a short-term measure. In cages for laying hens, it is sometimes possible to identify and remove the birds responsible, because in a small group such birds will have much better plumage than the others which they have pecked. They cannot be regrouped, however, because if they are they will resume pecking each other. Housing them singly is generally impractical, so the main option is to trim the beaks of these specific culprits. Beak trimming (see next section) does reduce the effect of feather pecking (Hughes and Michie, 1982), at least in the short term. In the long term, its effect is variable. In a study of a deep litter system, feather loss was reduced by beak trimming, while there was no corresponding effect of beak trimming in cages (Appleby *et al.*, 1988b).

9.9   Cannibalism

Cannibalism, like feather pecking, occurs on occasion among hens, turkeys, pheasants, quails and ducks. It sometimes follows on from feather pecking, for example when exposed skin is injured, but it more often arises independently. It is an emotive topic, but although the behaviour of one bird pecking flesh from another arouses emotion it is actually the behaviour of the pecked bird which is more difficult to explain. In hens, the most common form is vent pecking. One situation in which this starts is when a hen has just laid an egg in a nest box, facing inwards and the vagina is still partly everted. Other hens waiting to lay, investigating the nest box and the hen in it, peck at the soft, red vent area. If the skin is broken, here or elsewhere on the body, other birds then join in pecking, because these species of birds are attracted to blood. Further pecking and consumption of flesh then frequently result in death. Vent pecking also sometimes occurs in cage systems. The aspect which is least understood is that the pecked bird often makes surprisingly little effort to escape, despite the fact that it is likely to be in severe pain. Sometimes this may be because the pecked bird is a low ranking individual, as described above, which has been pecked aggressively so often that it has learned that it cannot escape.

Similarly, in a small cage or at high stocking density, birds may learn that they cannot avoid being pecked. Failure to attempt escape, then, may be due to 'learned helplessness', a state in which animals eventually become passive in reaction to suffering they cannot avoid (Seligman, 1975). However, freezing as a response to cannibalism occurs much more quickly than other, more definite examples of learned helplessness. One other possible explanation is that this freezing is related to the behaviour which occurs when one bird is having food pecked from its face by another (section 9.5). This would suggest that a bird is misled by circumstances (such as, perhaps, high stocking density) into performing an inappropriate and fatal behaviour pattern. In any event, it is evident that cannibalism is a major problem for both the animals concerned and the producer.

In contrast to feather pecking, cannibalism is more common in non-cage systems than in cages. This may be partly due to local crowding or social disturbance: in one deep litter house cannibalism began on a crowded, slatted area when some hens escaped from one pen into another (Appleby *et al.*, 1989). It will also be clear from the description above that there are particular pre-disposing factors, such as designs of nest box which encourage birds to face inwards and inadequate numbers of nest boxes for a flock of birds. However, the single most important factor is group size: in large groups there is more possibility of birds subsequently imitating each other's behaviour. As such, outbreaks of cannibalism are unpredictable in nature, occurring in some flocks but not in others and when they do occur mortality can be disturbingly high. Losses of up to 13% of a flock of laying hens have been reported in an aviary (Hill, 1986) and of up to 15% in both a strawyard (Gibson *et al.*, 1988) and a free range system (Keeling *et al.*, 1988). An outbreak of cannibalism can occur at any stage of the laying cycle. In the free range flock cited, cannibalism became severe after 11 months of lay, with most losses during the final 8 weeks (Keeling *et al.*, 1988). While cannibalism also occurs in cages, it does not spread so easily and the problem tends to be contained within the small groups.

Reduction of light intensity, if this is possible, can help to stop an outbreak of cannibalism and one other technique which can be combined with this is to introduce red lighting. This makes it more difficult for birds to see blood or wounds, while allowing them to feed and perform other behaviour. Neither of these methods is available in systems open to daylight, including strawyards.

Birds which are behaving as cannibals are sometimes seen with blood on their beaks, in which case they can be removed. This is more often possible in cages than in other systems, because a cannibalized bird in a cage only has a small number of flock-mates.

Beak trimming of chickens, which used to be called debeaking, is carried out to reduce aggressive pecking, feather pecking and cannibalism (Figure 9.9). It usually involves removal of between a third and a half of both the upper and lower mandibles; this is not just the horny beak but the underlying tissue as well,

a.

b.

Figure 9.9.   Two hens which were beak trimmed when young, in this case by removing part of the upper mandible. a. Partial regrowth of the mandible has occurred, producing a beak which looks fairly normal, but in which nerves may be abnormal. b. Here there has been malformation of the upper mandible and overgrowth of the lower manidble.

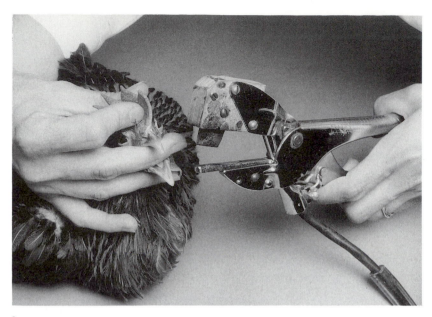

a.

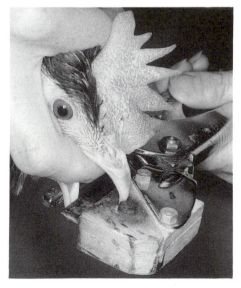

b.

Figure 9.10. Beak trimming, here seen in an adult hen. a. The tool has an upper blade which cuts down on to a bar; it is electrically heated to cauterize bleeding. The bird's tongue must be kept out of the way. b. Trimming of the lower mandible.

and use of a special cutting tool with a heated blade is common to cauterize the bleeding (Figure 9.10). It may be done shortly after hatching or at a later age, or at both if the beak regrows. It probably has two effects on pecking. First, it is likely that hens with trimmed beaks do not peck so strongly, for example in aggressive pecking, because to do so is painful. In fact, studies of the nerves in trimmed beaks suggest that trimming may result in long-term pain. Thus although the main reason for beak trimming is to prevent cannibalism, a welfare problem, the procedure is itself subject to criticism on welfare grounds (Gentle, 1986a). Second, trimming reduces the sharpness of the beak and the accuracy with which the bird can peck. Pecking frequency actually increases; for example, while feeding, more pecks are needed to achieve the same intake (Gentle *et al.*, 1982). Aggressive birds also peck subordinates more often, because the latter react less. However, the effect of pecks on the feathers of other birds, from either aggression or feather pecking, is reduced.

Beak trimming of large flocks is most practicable after hatching, but it is sometimes done at point of lay, before the pullets are moved to their laying accommodation. It is used as a preventative measure prior to housing in all systems, but more often for floor housing than for cages, as cannibalism tends to be worse in the former. It is usually effective in reducing the likelihood of cannibalism. In a comparison of deep litter with cages, outbreaks of cannibalism occurred in both systems in flocks which had not been beak trimmed, but not in flocks trimmed at one day old (Appleby *et al.*, 1988b). However, in the UK, it is now recommended that beak trimming should be carried out only as a last resort (MAFF, 1987a), because of the criticisms discussed above. This approach was adopted in a study of a perchery, which was divided into pens housing about 120 birds. The house was stocked with birds which had not been beak trimmed; beak trimming was carried out later on the birds in any given pen only if cannibalism occurred in that pen. As it turned out, beak trimming was necessary in five of the six pens in one year, but in no pens in the following year (Michie and Wilson, 1985). It must be recognized, however, that this approach would be impracticable in a large flock. In welfare terms, current systems housing large flocks of laying hens have two major, alternative problems: the chance of outbreaks of cannibalism and the effects on the birds of preventative beak trimming. These problems provide a major argument against such systems.

9.10 Causes of harmful social behaviour

The harmful behaviour patterns discussed above are often referred to as 'vices'. This term is also used to include other behaviour such as egg eating (section 11.8), behaviour which is inconvenient to the producer. It should be clear from the preceding sections that all these behaviour patterns are strongly affected by environmental factors. The term 'vice' is therefore inappropriate in that it suggests that the causes are mainly intrinsic to the birds: that the birds are 'to blame', rather than the environment. Further, it suggests that the problems associated with such behaviour are largely insoluble, because they are inherent. On the contrary, appropriate management techniques can often reduce the effects of such behaviour, when it occurs. Even more important, good management can help to prevent it occurring.

10

Reproduction

10.1 Summary

- This chapter is primarily about mating behaviour and its effects, with the exception of egg laying (Chapter 11). Artificial insemination and selection are considered and housing conditions and systems for breeding stock are compared.
- Under natural conditions galliforms form small harems, while ducks and geese are either monogamous or promiscuous. Courtship is important for successful mating.
- Selection for growth rate means that breeding birds of meat strains have to be food restricted to maintain good fertility. Breeding groups can be kept in cages, but the commonest housing method is in pens, often of a very large size.
- Male fowls, unlike females, are strongly sexually motivated and mate frequently, preferring a series of different hens. In commercial conditions one cockerel per 10–16 hens gives good results, whereas for ducks one drake per 5–8 females is needed.
- Hens typically mate about once a day, but artificial insemination at intervals of 1–2 weeks is sufficient to maintain maximum fertility, at levels of 85–90%.
- Mating efficiency in a flock usually decreases over time; to counter this younger males may be added about halfway through the laying year. Semen quality and volume are governed by several factors, including housing conditions and stress.

10.2   Natural behaviour

The usual mating system in galliforms is the harem, with one male monopolizing mating in a group of females throughout the mating season. Other males are sometimes tolerated near harems, or they may be solitary or form separate groups. Jungle fowl, living in dense vegetation, seem to be territorial, with small harems of up to four hens (Collias and Collias, 1967). In more open conditions, they form larger groups (Collias *et al.*, 1966). Studies of feral fowl have described harems of four to twelve hens (McBride *et al.*, 1969), but these must also depend on precise local conditions. Ducks and geese are more variable in their reproductive behaviour: both groups are monogamous in certain conditions, but mallards will often mate promiscuously even in the wild and geese will do so under domestication.

In all poultry, the male usually takes the initiative in mating. In the wild this is restricted to a breeding season in spring, but there is no limited receptive period in females comparable to oestrus in mammals. However, a female will usually only be receptive after courtship by a male, which varies between species and so ensures that the offspring will be viable. In most species, female receptivity involves crouching for the male to mount, although this does not occur in quail (Ottinger and Brinkley, 1979). In jungle fowl and domestic fowl, courtship is quite complex in its full form, with a chain of stimulus–response patterns between male and female (Fischer, 1975). The male's courtship behaviour has been well described by Wood-Gush (1971). Early in courtship, the male performs some actions which also occur in aggression between males, such as wing-flapping and waltzing; the latter is a sideways or circling movement with one wing trailing. These may intimidate the hen and encourage her to crouch. If she moves away, though, a behaviour pattern called cornering may be used. Cornering usually involves stamping and calling in a corner and often attracts the hen to approach. This may be partly because it is sometimes associated with nest site selection by the hen and in fact mating often follows nesting (McBride *et al.*, 1969). Another common behaviour which occurs late in courtship is the rear approach. This is, as it sounds, often an immediate precursor of mounting (Wood-Gush, 1971). Similar sequences of behaviour between male and female occur in turkeys (Figure 10.1) and other poultry.

A female would normally lay a clutch of eggs, of a size which varies among species, then become broody and begin to incubate them. Laying can be extended in certain wild birds by the removal of eggs, but not indefinitely: most selection for egg production has consisted of the extension of this process and hence the avoidance of broodiness (section 10.7). The behaviour associated with egg laying will be considered in Chapter 11.

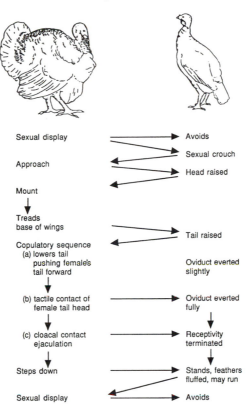

Sexual display	→	Avoids
	←	Sexual crouch
Approach		Head raised
	←	
Mount		
↓		
Treads base of wings	→	Tail raised
	←	
Copulatory sequence (a) lowers tail pushing female's tail forward		Oviduct everted slightly
↓		↓
(b) tactile contact of female tail head	→	Oviduct everted fully
↓		↓
(c) cloacal contact ejaculation	→	Receptivity terminated
↓		↓
Steps down	→	Stands, feathers fluffed, may run
Sexual display	← →	Avoids

Figure 10.1. A schematic illustration of the sequence of reproductive behaviour in turkeys (Hale *et al.*, 1975).

10.3   Sexual development

Birds must learn the characteristics of appropriate mates for normal sexual activity: the fact that this knowledge is not innate is demonstrated by the observation that hand-reared birds often show courtship or sexual crouching to humans. Learning is most likely to occur during a sensitive period which precedes sexual maturity. This period is poorly defined, but in male chickens it is at about 10 to 12 weeks old (Siegel and Siegel, 1964). Until recently it was common to rear male and female broiler breeder chickens separately for independent control of body weight and this probably resulted in decreased mating behaviour and fertility (Wood-Gush and Osborne, 1956). In broiler breeders it is now common to rear the sexes together, since the development of a simple method for feeding them separately once they are adult (section 8.8). In ducks, sexual imprinting is important and so breeding ducks are also reared with the sexes together (Hearn and Gooderham, 1988). Turkeys are still reared separately, because, with artificial insemination prevalent, mating behaviour is not important. It was also thought that there was no deleterious effect on semen production (Wood-Gush and Osborne, 1956; Siegel, 1965), but while this may be true for semen volume, recent studies suggest that semen quality is poorer in males isolated from females (Jones and Leighton, 1987).

Rapid growth and large appetite are desirable traits in broilers, turkeys, ducks and geese being grown for their meat. However, the same traits are also expressed in the parent stock from which the meat birds are derived and in these breeding birds they can cause problems. Fast growth is associated with low fertility in females (Soller and Rappaport, 1971). In addition, excess body fat depresses reproductive activity (Lorenz, 1959) and large breeding birds would cost more to feed. Breeders are therefore fed only sufficient during rearing to reach about 50% of their potential body weight. There is increasing evidence, however, that in the simple housing conditions used for these flocks such restriction is stressful. For example, broiler breeder hens on restricted diets show the abnormal behaviour of spot-pecking: stereotypic pecking over long periods at small marks on the pen wall or elsewhere (Savory, 1989a). Males are also restricted in body weight, even though restriction may delay sexual maturity (Jones *et al.*, 1967). This is partly because full-size males would not be suitable for small females and also because males fed *ad libitum* become obese and suffer from foot and skeletal problems. These make them unwilling or unable to mate under natural mating conditions.

In chickens, one alternative to restricting the food of breeders has been the development of dwarf strains. These depend on a sex-linked recessive gene: a gene which results in small body size, but only in females, and which is masked in the offspring which these females produce with normal males (Figure 10.2). The offspring are, to all appearances, normal broilers, but the problems of rapid growth or food restriction in their mothers are avoided. These strains have been widely adopted in France, but to a lesser extent in other countries. This is partly because they require careful management; for example, the mating between dwarf females and normal males can cause problems. However, there have been almost no formal studies of the behaviour of these strains. It is also still not proven that dwarfing is fully recessive: some measurements suggest that the resulting broilers do not grow as quickly as others. One explanation for this could be that egg size from dwarf females is slightly reduced.

Attempts to assess the sexual potential of males very early in life have not been successful (Wood-Gush, 1963a), but tests of libido soon after sexual maturity give quite good predictions of fertility of subsequent mating (McDaniel and Craig, 1959). It may therefore be possible to select suitable males before they are used for breeding (Justice *et al.*, 1962).

In contrast to many other groups of birds, in poultry the presence of males is not necessary for sexual maturity of females. It is on this characteristic, of course, that the egg industry depends, since hens (and turkeys) will lay large

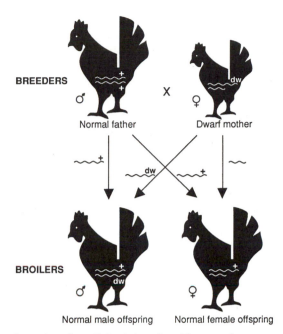

Figure 10.2. Genetics of sex-linked dwarfing. The sex chromosomes are shown, which in birds are similar in males, dissimilar in females. In the female broiler breeder the Dwarf gene (dw) is the only copy and results in dwarfing. In the male broiler the Dwarf gene is paired with a normal gene (+) from the father; it is recessive so its effect is masked. The female broiler gets its only copy of the gene from the father: a normal gene.

numbers of eggs in the absence of males. There is evidence in turkeys, though, that egg production is increased by male stimulation (Jones and Leighton, 1987), so it seems that housing the sexes separately is disadvantageous for both female and male breeding potential. Similarly, in quail the onset of lay is advanced if females can hear male vocalizations (Guyomarc'h *et al.*, 1981).

10.4 Sexual motivation

It is sometimes suggested that depriving females of sexual contact is a welfare problem. In fact, while sexual motivation in males is strong and can be measured experimentally, there is little indication of strong motivation for mating in female poultry (Duncan and Kite, 1987). On occasion a female which has not mated for some time will approach a male, but males usually take the initiative in copulation itself and male libido is one of the main determinants of mating success. Muscovy ducks are an exception in this; female muscovies will readily approach males for mating (Raud, 1990).

Individual males vary in libido, which suggests the possibilities of choosing the most effective males, as mentioned above (Justice *et al.*, 1962) and of selecting strains for mating behaviour. Such selection has been successful experimentally, but requires careful identification of the behaviour to be selected. Thus in one study which recorded overall mating frequency, a line which mated more frequently than controls was produced. However, many of these matings were incomplete (Wood-Gush, 1960). Another study took more detailed observations and selected on the basis of complete matings only, with greater long-term success (Siegel, 1965). Unfortunately, selection for frequent mating tends to result in low semen volume. It may be possible to control for this while still increasing libido, but this has not been done commercially.

A male that has mated with a female may do so again if she does not move away and continues to be receptive. However, repeat matings with one female only occur after increasing intervals of time. It seems likely that such matings are advantageous in that they increase fertility, but that successive matings yield a diminishing advantage. The decline in motivation of the male is not caused by fatigue, but by habituation to the stimulus female: the male will resume active mating if another receptive female is available. This is called the 'Coolidge effect', after a well-publicized occasion on which it was explained to the American president and his wife. Habituation to particular females cannot be a problem in large breeding flocks. It is not known whether it can occur in small breeding pens or cages.

In females, there is variation between individuals and between species in how quickly they will mate again. Turkeys are more likely than hens to avoid repeat mating and may do so for some days. This even occurs after incomplete copulation or after mounting by another female (Hale, 1955), which must certainly reduce fertility.

Even in large flocks there is, of course, a limit to the number and frequency of copulations by both males and females. Each sex does, however, alter its behaviour to compensate to some extent for satiation in the other. Classic experiments by Guhl (1953) showed that cockerels would be more active in courtship when hens were satiated and that hens would crouch more readily to satiated cockerels (Figure 10.3).

10.5   Mating

Many farmyard flocks and small-scale operations with chickens, ducks or geese have groups of one male with a number of females which are similar to the natural harems of galliforms. The nearest equivalent to these in large-scale poultry production are the cages now sometimes used for breeding groups of chickens, particularly where selection is being carried out. In these, one cockerel is typically housed with eight to ten hens, although some trials have used two cockerels with a larger group of hens and satisfactory results have been obtained for

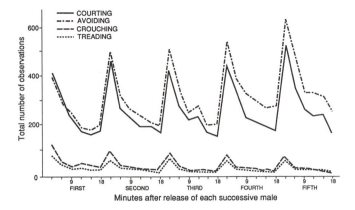

Figure 10.3. Compensatory sexual behaviour. Five males were introduced one at a time, for 18 min each, into a pen of hens. Hens showed satiation by crouching less for each successive male. Males compensated by courting more, and thus each achieved similar success in treading. Each soon became satiated, though, and courting rapidly declined (Guhl, 1953).

fertility (Campos *et al.*, 1971, 1973). One study on the details of cage design concluded that firmness of flooring was important for successful mating but came to no conclusions on other features (Bhagwat and Craig, 1975). No studies of mating behaviour in cages have been published, though. It is not even known whether, in a cage with two males, both are active in mating or one predominates.

It is most common, however, for breeding fowl and ducks to be housed in large floor pens. In ducks, a group size of about 400 is recommended (Hearn and Gooderham, 1988) but breeding houses for chickens usually contain 4000 to 8000 birds. Commonly used sex ratios for such flocks are given in Table 10.1. In some cases, these are the result of research; thus, studies on flocks of New Hampshire chickens found maximum fertility with 6 or 7 cocks to 100 hens (Parker and Bernier, 1950). However, the mating success of a flock will be affected not just by the sex ratio but also by the precise housing conditions and

Table 10.1. Sex ratios in breeding flocks.

Species	Type	Sex ratio Females/male	Males/100 females
Chickens	Egg-type breeders	14–16	6–7
	Broiler breeders	10–12	9
Ducks	Egg strains	8	12–13
	Meat strains	5–6	16–20

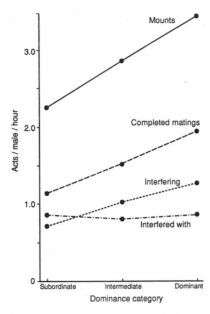

Figure 10.4.   Effect of rank on mating behaviour of White Leghorn cocks. Males frequently interfered with each other's mating (Kratzer and Craig, 1980).

the behaviour of the birds. Males often interfere with the mating of others, particularly with those subordinate to themselves (Figure 10.4). Furthermore, in certain conditions mating of low-ranking males can become completely suppressed, resulting in what has been called 'psychological castration' (Guhl *et al.*, 1945). These conditions are not known in detail but probably involve small groups and crowding, which encourage a strong hierarchy among the males. High stocking density, in fact, directly restricts courtship and mating (Kratzer and Craig, 1980). However, this is not necessarily reflected in reduced fertility, at least in cages (Bhagwat and Craig, 1975).

In larger groups, rank seems to have less effect on mating by different males (Craig *et al.*, 1977; Kratzer and Craig, 1980). In very large groups, it used to be thought that birds would form sub-groups (McBride and Foenander, 1962) which might act in a similar way to harems, but actually both males and females wander widely over most or all of the area (Appleby *et al.*, 1985). It is possible that this leads to less interference in mating than in smaller groups. However, in any conditions, variation in the sexual activity of individual males is likely to mean that the effective sex ratio is different from the actual sex ratio. Little is known about these aspects of reproduction, though, because there have been almost no systematic studies of mating in commercial conditions.

Similarly, there have been few studies published of individual variation in mating frequency of male or female poultry in flocks and few even of average frequencies. One study of chickens reported males mating up to 41 times per day (Parker *et al.*, 1940) but that figure was atypical. A study of pens housing about 150 hens estimated that males mated about 5 times daily on average and that this did not increase even with a sex ratio of 24 females per male (Craig *et al.*, 1977). For females, a similar study in pens suggested they mated more than once daily (Kratzer and Craig, 1980), while a small-scale study in a commercial broiler breeder house found an average of 0.48 matings per day (Appleby and Cunningham, unpublished observations). These frequencies seem to be higher than necessary, because fertility is just as high when artificial insemination is performed at weekly intervals (Craig, 1981). However, as with males, there is variation between individual females and average mating frequency is not a very useful measure. For turkeys, carrying out artificial insemination at 2 week intervals is sufficient for maximum fertility early in lay, although inseminating at weekly intervals is more common practice (Clayton *et al.*, 1985).

Mating frequency declines with age (Figure 10.5), but this does not result directly in lower fertility, perhaps because, as just suggested, early mating is more frequent than necessary. The decline in fertility which occurs during the laying year is, however, associated with other age-related changes in sexual behaviour (Duncan *et al.*, 1990; see next section). In broiler breeder flocks, there is a practice, called 'spiking', in which young males are added to the flock

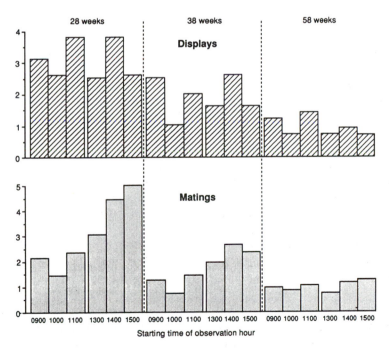

Figure 10.5. Effect of age on mating behaviour of broiler breeder males. Both displays and matings declined with age, at all times of day (Duncan *et al.*, 1990).

about mid-way through the laying year in an attempt to reduce this decline. It is believed that they increase the mating frequency both directly, by themselves mating, and also indirectly, by stimulating the activity of the resident, older males. No evidence has been published on the effectiveness of this practice, but in one study of such a house the new, young males carried out a higher proportion of matings than would have been expected from their numbers (Figure 10.6). This practice is more common in Europe than in North America. Males may also have to be added to breeding flocks if many of them die or have to be culled, so changing the sex ratio.

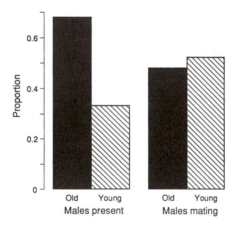

Figure 10.6.   'Spiking' in a commercial broiler breeder flock. Three groups of young males, about 20 weeks old, were added when the original birds were 43, 46 and 54 weeks old. Observations were made when females and 'old' males were 58 weeks of age (Appleby and Cunningham, unpublished observations).

There is some evidence that the proficiency of mating by males increases with time. In a study of changes which occurred after the introduction of cocks to flocks of hens, the proportion of mountings which led to complete matings increased steadily from 37% initially to 69% by the fifth week. However, most of this change resulted from a decline in mating interference and in incomplete matings. The actual number of complete matings did not increase after the second week (Figure 10.7).

Distribution of mating through the day is affected by the egg-laying cycle, because fertility is less around the time of oviposition. Hens lay mostly in the morning. Correspondingly, they mate more frequently in the afternoon than in the morning (Upp, 1928; Craig and Bhagwat, 1974), although there is also some evidence of a peak in sexual activity after lights-on. Quails, by contrast, lay in the afternoon and mate most in the morning and evening (Ottinger *et al.*, 1982).

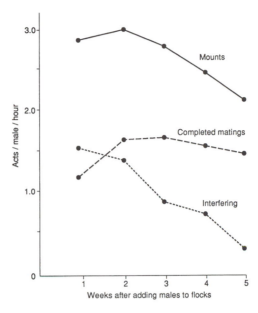

Figure 10.7. Changes in mating behaviour of White Leghorn cocks with time. The proportion of successful mounts increased from 37 to 69% in five weeks (Kratzer and Craig, 1980).

10.6   Fertility and hatchability

The number of offspring produced by a breeding flock is the result of production of eggs suitable for incubation (i.e. clean, uncracked eggs of acceptable size, shape and shell quality), fertility (the proportion of eggs which are fertile) and hatchability (the proportion of incubated eggs which hatch). Fertility is affected by many factors (see Lake, 1969) and it will be clear from the previous section that these include behaviour and management. Hatchability is also influenced, to a lesser extent, by behaviour and management.

In breeding fowl, fertility is often 95% or higher early in the laying year, but declines sharply later, particularly after 50 weeks of age (Kirk *et al.*, 1980). This decline has been attributed to males rather than females, because fertility can be maintained by artificial insemination (Brillard and McDaniel, 1986). Duncan and colleagues (1990) therefore studied changes in the behaviour of male broiler breeders with age. They found, as mentioned in the previous section, that libido and mating declined with age, but that these changes were not accompanied by decreased fertility in all groups. In particular, fertility remained high in groups where males were slightly food-restricted. It decreased, though, with males fed *ad libitum* (Figure 10.8) and the authors suggest that this was an effect of male bulk or conformation interfering with semen transfer during copulation. In other words, some matings by these birds which appeared to be complete were in fact probably not effective. Other reports also suggest that excessive weight in males reduces mating efficiency (McDaniel and Craig, 1959; Rappaport and Soller, 1966) and this problem is now usually avoided by feeding males and females separately (section 8.8).

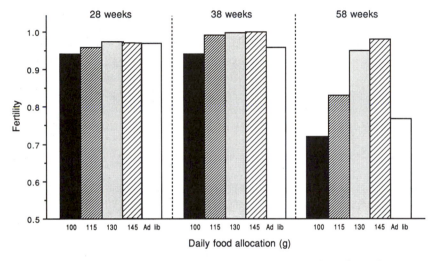

Figure 10.8.   Effects of food allowance on fertility in broiler breeders. The proportion of fertile eggs declined with age in flocks where males were severely restricted, or not restricted at all, but not where they were slightly restricted (Duncan *et al.*, 1990).

It is certainly the case in turkeys that selection for body size has reduced fertility of natural mating, to the extent that it is no longer viable commercially. Artificial insemination in turkeys generally results in fertility over 85%; with proficient technique, 95% can be achieved in the first half of the laying period and 90% in the second half (Clayton *et al.*, 1985). Artificial insemination is also the rule in guinea fowl. In other species of poultry, it is not generally economic, although it has been quite widely used for chickens in Israel and Japan (Cooper, 1969). It is also used for special breeding programmes or to perpetuate highly inbred lines with poor fertility for genetical studies (Lake, 1969).

Knowledge of behaviour can also be useful in artificial insemination. One clear example is given by the fact that daily variation in insemination success closely matches the variation seen in mating frequency. In turkeys, this has been explained by the fact that contractions in the oviduct around the time of oviposition obstruct artificial insemination just as they do natural fertilization (Brillard *et al.*, 1987). This would suggest, then, that artificial insemination of turkeys and chickens should be done in the afternoon, while that of quail should be done in the morning. Fertility by artificial insemination can also be influenced by social factors, such as the presence of males in the cage house (Ottinger and Mench, 1989).

Housing conditions also influence semen production in males. Perhaps surprisingly, both cockerels and stag turkeys produce greater volumes of semen when they are kept in cages rather than pens (Siegel and Beane, 1963; Woodard and Abplanalp, 1967). However, this may actually be an effect of group size, with male–male mounting in groups reducing the remaining semen volume compared to isolated males (Ottinger and Mench, 1989). When males are housed in cages it is important to have good conditions, as emphasized by Cooper (1969), who stresses that cage height should be sufficient to prevent the comb touching the roof and that flooring should be designed to prevent foot damage in heavy males. Crowding and large group sizes should be avoided, because stress can depress reproductive ability (Ottinger and Mench, 1989). Separation of the sexes also affects semen quality, as already mentioned: quality is poorer in non-mating turkey males than in those allowed to mate naturally (Jones and Leighton, 1987).

Fertility is assessed by candling eggs – placing them over a bright light – during the early stages of incubation, often after about 7 days. During this process, fertile eggs appear opaque because of the development taking place, while infertile eggs which are not developing appear clear. Those which are fertile are taken on to further incubation, so the later measure of hatchability reflects embryo survival between laying and hatching. Several of the factors which affect fertility also affect hatchability, including certain aspects of behaviour and management. For example, hatchability will be reduced for floor eggs and eggs laid in dirty nest boxes and hatchability varies for eggs laid at different times of day. It also declines with time after insemination, especially towards the end of the fertile period (Landauer, 1967). The mechanisms for these latter two effects are not known (see Lake, 1969).

10.7   Selection

The main traits selected for in the poultry industry are those relevant to productivity, such as egg production and efficient growth, and considerable changes have been achieved in these traits (Chapter 2). Insufficient emphasis, though, is placed on the fact that many of these traits have behavioural components, including components of reproductive behaviour. An understanding of behaviour may help in the improvement of productivity in the future, while also helping to reduce associated welfare problems.

One of the clearest examples of the involvement of behaviour in productivity is that early selection for egg production in domestic poultry consisted primarily of selection against broodiness. The incidence of broodiness in chickens was reduced from over 90% to less than 20% after five generations of selection, as

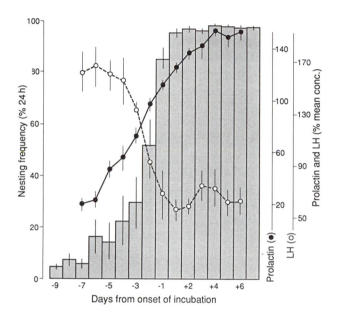

Figure 10.9.   Hormonal changes and broodiness in bantam hens. The hormones prolactin and luteinizing hormone (LH) are both produced by the pituitary gland at the base of the brain; LH acts on the ovary to cause ovulation. An increase in production of prolactin is followed by a decrease in LH and by the start of incubation (Lea *et al.*, 1981).

early as 1920 (Goodale *et al.*, 1920). Commercial selection against broodiness, however, was an indirect result of that in favour of higher egg numbers. Broodiness still occurs occasionally in laying hens, more often in broiler strains (in up to 10% of birds) and inconveniently often in turkeys, where it can affect up to 50% of the flock (Harvey and Bedrak, 1984). There are also indications that it is currently increasing again in floor-housed birds. It seems likely that direct observation of the behaviour of birds under selection could have reduced this problem further in the past and that it could be important in doing so again in the future. Understanding of the genetic and endocrinological basis for broodiness is now advancing (Figure 10.9) and should also improve its control in future. Broodiness is a sex-linked characteristic, which has the surprising effect that whether a particular bird becomes broody or not depends more on the genes it has inherited from its father than on those from its mother (Figure 10.10; Craig, 1981).

Another problem for egg production which can be identified by observation of behaviour is that of internal laying (section 11.3). This can be distinguished from failure to ovulate by the fact that pre-laying behaviour occurs, even though there is no egg to lay, so it is possible that with appropriate records of behaviour this condition might be selected against.

There is also an increasing interest in selection for behavioural traits which are relevant to welfare (section 7.2). These include aspects of reproductive

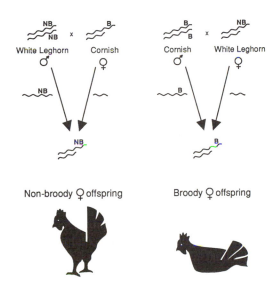

Figure 10.10. Genetics of broodiness, illustrated by crosses between two breeds. Most of the genes which affect broodiness are on the sex chromosome of which the female only has one copy. Female offspring therefore obtain the relevant genes from their father. In White Leghorns broodiness is rare, so 'non-broody genes' (NB) are common. In the Cornish breed broodiness is common, caused by 'broody genes' (B) (after Craig, 1981).

behaviour, such as pre-laying pacing (section 11.6). Studies on selecting behaviour and traits relevant to welfare form part of a wider debate on whether poultry can be adapted genetically to suit commercial environments, or whether the main emphasis should be on providing appropriate environments for the birds concerned (see Conclusions).

10.8   Sexual behaviour and management

It has been a regular theme of this chapter that little is known in detail about appropriate management of sexual behaviour. In particular, there have been almost no studies of sexual behaviour in commercial conditions. This is surprising in view of the fact that selection of birds for mating and management of sexual behaviour have far-reaching importance both for the birds and for their economic performance.

11

Egg laying

11.1 Summary

- Under natural conditions female poultry leave the flock to lay in a secluded nest site, often enclosed and usually a shallow scrape or sometimes bowl shaped. Once the clutch is complete and incubation has begun the broody bird only leaves the nest for about one hour per day.
- Pre-laying behaviour is controlled by the ovulation, which occurs about 24 h previously, and the release of hormones from the post-ovulatory follicle. Birds showing pre-laying behaviour are strongly motivated to find a suitable nest site. Problems in finding nest sites can result in increased aggression, both in cages and non-cage systems.
- Egg laying is closely related to the timing of the photoperiod; in 24 h light–dark cycles, hens lay almost all eggs within 6 h of lights-on. However, on 28 h ahemeral cycles eggs are mostly laid during darkness.
- Eggs laid on the floor can be a major problem in non-cage systems; to reduce them requires attention to rearing and housing conditions, nest-box design and management. With hens, increasing mobility by giving young birds access to perches and experience of nest sites before point-of-lay, are both helpful.
- Hens will lay satisfactorily in a wide variety of nests and in most systems these are of a rollaway type. If available, however, hens prefer nests with litter. About one nest per four to five hens is ideal. Too few can result in floor laying.
- Agitated pre-laying pacing, though sometimes seen in non-cage systems, is commonest in cages and generally more severe in light than in medium hybrid hens. It is the most serious behavioural problem in cages.
- The number of eggs laid is similar in all systems, but more are lost in

175

non-cage systems. In most, egg production declines with increased stocking density.

11.2   Natural behaviour

In feral or wild poultry, a female which is about to lay an egg leaves the social group and normal home range and moves away to choose a nest site or to find the site chosen on a previous day. In domestic fowl, the male sometimes accompanies the female and it has been suggested that both are involved in nest site selection (McBride *et al.*, 1969). An alternative suggestion, though, is that the male's interest in nesting is because mating is more effective after laying than before (section 10.5) and in fact often occurs shortly after the female leaves the nest. Nest sites are usually well defined, in places such as the foot of a slope or under a bush, secluded from disturbance and enclosed for protection from predators. If loose material is available it is shaped into a hollow nest bowl, but poultry will also nest in a hollow scraped in the earth or on the flat ground (Duncan *et al.*, 1978).

After an egg is laid, the female sits on it for only a short while before returning to normal behaviour. This remains true even when several eggs have accumulated in the nest. When the clutch is complete, however, and the female becomes broody, incubation is almost continuous. In most poultry, a broody bird only leaves the nest for about one hour each day to feed, drink and defecate; quail, however, leave the nest several times daily. Only when incubation starts do the embryos begin to develop, so development is synchronous even in eggs which were laid days or weeks apart. It is important that hatching of all eggs occurs over a short period because chicks are mobile soon after hatching and a brood hatched at different times would become separated. Maternal behaviour, imprinting and socialization are considered in section 9.3.

11.3   Control of egg laying

Much of the behaviour associated with egg laying is under genetic control. This may contribute to the very consistent, even rigid, way in which this behaviour is expressed (section 7.3), although details are modified by the environment, including the housing system. Some features of nesting are even alike between different species of poultry.

Genetic effects are also evident from differences between strains in the details of pre-laying behaviour (section 11.6) and from the reduction in broodiness caused by selection for egg production (section 10.7). There may have been incidental selection in the past for or against certain aspects of laying behaviour. For example, in farmyard flocks over many centuries those birds nesting in inaccessible sites were more likely to rear a brood than those laying in nest boxes, in which the eggs were taken away from them. Conversely, in more modern systems where breeding birds are enclosed in houses with nest boxes there is indirect selection against floor laying. Breeding programmes use trap nests to identify which birds lay which eggs: these have a lever over the entrance which triggers a door to close when a bird enters. The door remains shut until it is opened manually and the egg labelled. Eggs laid on the floor cannot be attributed to the birds which laid them and these birds are therefore likely to be culled as poor layers. Even without trap nesting, floor eggs are more likely to be broken or dirty than nest eggs and therefore are rejected for incubation. Such indirect selection can have only a minor effect, however, because the causes of floor laying are complex (section 11.5).

While the actual genetic mechanisms which control nesting are not fully understood, a considerable amount is known about physiological control. It might be expected that pre-laying behaviour would be set off by the presence of an egg in the shell gland, ready to be laid. In fact, it is triggered by ovulation, approximately 24 h earlier (Wood-Gush and Gilbert, 1964) and the release of oestrogen and progesterone from the follicle after ovulation (Wood-Gush and Gilbert, 1973). These hormones act on the central nervous system and cause nesting behaviour to be initiated after a suitable time interval. Meanwhile, the egg is developing quite independently of this process (section 1.15) and it is ready to be laid by the time nesting behaviour has begun. Pre-laying behaviour and oviposition are therefore usually appropriately synchronized (Figure 11.1).

One consequence of this mechanism, however, is that once ovulation has occurred, pre-laying behaviour will proceed even if something goes wrong with normal development of the egg: it will start at the expected time but without an egg to be laid. The most common cause is 'internal laying', where the ovum is not picked up by the oviduct and enters the peritoneal cavity where it is resorbed internally; alternatively, sometimes an egg is laid prematurely without a hard shell (Wood-Gush, 1963b). These problems occur at quite a high frequency in all systems, but are largely unrecognized because behaviour is not usually recorded. One study which did record nesting behaviour using trap nests, though, suggested that up to 12% of potential eggs were being lost by internal laying (Wood-Gush and Gilbert, 1970). It would certainly be possible to investigate these problems further in commercial strains by studying the frequency of nesting without laying.

Once pre-laying behaviour has been triggered, birds have very strong motivation to find a suitable place for laying (Duncan and Kite, 1987). What constitutes a suitable place and how this is reflected in the behaviour of the birds are discussed in section 11.5.

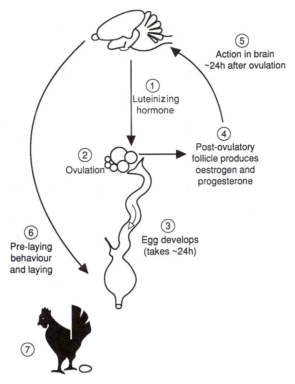

Figure 11.1.   Control of pre-laying behaviour. Ovulation in the single, left ovary is under endocrine control. While the egg develops in the oviduct there is then hormonal feedback on the brain, which triggers the behaviour at an appropriate time. See also section 1.14.

11.4   Timing of egg laying

The fact that ovulation occurs around dawn and oviposition about 24 h later (except in quail, which lay in the afternoon) means that the timing of egg laying is strongly influenced by the ambient light in open systems and by the lighting regime in closed houses. For laying hens in closed houses it is usual to start the light period at 03.00 h or even earlier, so that most eggs have been laid before the beginning of the operatives' working day. This does have risks, however; for example, birds need to be fed at a similarly early time and feeding systems often run unsupervised. Faults may not be detected for some time, leading to problems for both the birds and the owner. The situation becomes more compli-cated in some of the artificial regimes which are used for laying hens (section 3.9). With ahemeral cycles where the day and night add up to more than 24 h, the delay between ovulation and laying is longer than normal but still short enough that laying occurs mostly in darkness. If such cycles are used in floor-housed flocks with nest boxes, floor laying increases greatly, so they tend to be restricted to caged birds. With some intermittent lighting schedules, there is no distinct dawn and so laying occurs over a much longer period.

By the mechanism of triggering by ovulation, pre-laying behaviour can only occur during a certain period (Figure 11.1). If oviposition is delayed beyond this period, it is not accompanied by pre-laying behaviour and the egg is laid in the course of other activity, often with hardly even a change of posture. The most common cause of such a delay is social interference between birds. This may occur where there are nest boxes, but they are all occupied, particularly if high-ranking birds defend one or more boxes against those of low rank (Perry *et al.*, 1971). To prevent floor laying it is important to provide sufficient nest boxes (Appleby, 1984). Delay of oviposition may also be caused by human disturbance during nesting.

It is also possible that some management practices contribute to delay of egg laying, particularly feeding birds during peak laying time. When limited food is provided on a schedule, feeding behaviour of hungry birds sometimes suppresses their pre-laying behaviour and restricted-fed birds are sometimes seen to lay while feeding. In one study on broiler breeders, however, varying feeding time had no effect on the proportion of eggs laid on the floor (Hearn, 1981).

11.5   Nest-site selection

All housing systems for laying poultry, except battery cages for hens, involve the collection of eggs from nest boxes. The behaviour of birds choosing where to lay is therefore critical in such systems. It also has important effects in cages, since eggs laid at the rear of the cage are more likely to be cracked as they roll forward than those laid at the front. In cages with perches, laying from the perch is another cause of cracked eggs (Duncan *et al.*, 1992). No remedy for these problems is currently apparent, although addition of nest boxes to cages is being studied (Appleby, 1990).

Failure to use nest boxes can be a major economic problem in most or all non-cage systems and also potentially in modified cages (Appleby, 1984; Hill, 1986). In severe cases, 50% or more of eggs are laid on the floor. These are labour intensive to collect and attempts to prevent floor laying also involve much work. Floor eggs are often broken, which encourages egg eating, or dirty, which reduces value or hatchability (Hodgetts, 1981). Floor laying is variable, not just between systems but within systems, even from flock to flock or from pen to pen and this variation has in the past seemed intractably unpredictable. However, there is increasing understanding of the factors which affect nest-site selection and floor laying. These include rearing conditions, housing conditions, nest box design and management, and human intervention (Appleby, 1984).

The conditions in which birds are reared affect later nest-site selection in two ways. First, they affect development of mobility, which has important effects in many poultry on their ability to gain access to nest boxes. In particular, some strains of hens, including at least medium hybrid layers and broiler breeders, learn most readily to jump or flap from the ground to higher levels if they have the opportunity to do so when they are young (Figure 9.1; Faure and Jones, 1982; Appleby and Duncan, 1989). Most nest boxes for adult hens are raised above ground level, so hens reared with no experience of perching, either in cages or on litter, lay many floor eggs as adults (Craig, 1980; Appleby *et al.*, 1986). In some rearing houses the problem is further exacerbated by the use of electric wires to prevent birds sitting on feeders, effectively training them not to perch. Avoidance of such mistakes and provision of perches during rearing, can greatly reduce later incidence of floor laying (Figure 11.2; Appleby *et al.*, 1983, 1988a; Brake, 1987). Second, rearing conditions may influence later nest-site preferences. Most such effects are likely to be subtle; for example, experience of dark or bright conditions affects choice of dark or bright nest boxes (Wood-Gush and Murphy, 1970; Appleby *et al.*, 1984b). However, it has also been suggested that allowing birds to investigate boxes before they are mature increases their readiness to use them later (Rietveld-Piepers *et al.*, 1985). Certainly floor laying is worse if birds are kept in rearing houses until after maturity and establish laying patterns in the absence of boxes (Dorminey, 1974).

Conditions in adult laying houses also affect mobility of birds and thus their use of nest boxes. In deep litter houses for laying hens, if drinkers or roosting areas are raised above ground level this encourages birds to perch (Appleby,

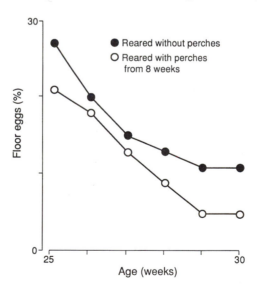

Figure 11.2.   Effect of providing perches during rearing on floor laying in commercial flocks of broiler grandparents. In the experimental flock perches were provided later than ideal, 8 weeks after hatching, but this still considerably reduced floor laying with raised nest boxes (Appleby *et al.*, 1988a).

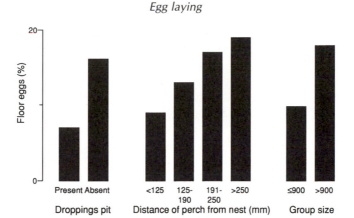

Figure 11.3.   Effect of housing conditions on floor laying: average results from a survey of 53 commercial pens of broiler breeders (after Brocklehurst, 1975).

1984) and increases use of raised nests (Figure 11.3; Maguire, 1986). On the other hand, floor laying is a major problem in some aviaries (Hill, 1986) and percheries (Anon., 1983), even though hens in these systems must be able to perch. It is possible that other factors are affecting the accessibility of nests in these houses. Other aspects of housing conditions which have been suggested as influences on floor laying include lighting, floor material and temperature, but evidence for these effects is equivocal (Appleby, 1984).

The fact that mobility of at least some types of poultry is restricted means that accessibility of nest boxes is one of their most important features. Ground level boxes may be advantageous, although some birds may even have difficulty using these if they have to step up into them (Appleby *et al.*, 1988a). With raised boxes for hens, easy access from an alighting rail or similar arrangement is essential (Figure 11.3; Appleby, 1984). In fact, the high proportion of hens which do use raised boxes is quite surprising, since hens and related birds nest on the ground in the wild. A likely explanation is that the main characteristic such birds use to select a nest site is enclosure, or protection. Nest boxes are more enclosed than any natural sites and so are usually chosen in preference to positions on the floor (Appleby and McRae, 1986). If there are enclosed sites on the floor, however, in corners or under nest boxes, these may be as attractive as the boxes themselves. Houses should be arranged to avoid providing such sites. The requirement for enclosure is not stringent and most designs of nest boxes are sufficiently enclosed. Indeed, the fact that many different designs of nests have been successful (Figure 11.4) suggests that important features are quite simple (Smith and Dun, 1983).

Features often thought to be important such as darkness and seclusion may affect which nest boxes birds choose (Figure 11.5), but are unlikely to affect the choice between nest boxes and the floor. Similarly, although birds prefer some nesting materials to others (Huber *et al.*, 1985) and prefer nests containing eggs

Figure 11.4.   Some common types of nest box. From left to right: wooden boxes with wood shavings as litter, metal boxes with plastic rollaways, wooden boxes with 'astroturf' rollaways and metal 'autonests' with buckwheat litter on a conveyor (Appleby *et al.*, 1988c).

Figure 11.5.   Traditional ideas about poultry behaviour are not always right. In this experiment, some hens chose the dark nest box, as expected, but others chose the light one (see Appleby *et al.*, 1984b).

to others (Kite *et al.*, 1980), there is no evidence that these factors influence floor laying. Only if there is no nesting material at all does floor laying seem to be worse. This is particularly important with regard to automatic egg collection systems for hens. Rollaway nests, which allow eggs to roll into a collection channel, are sometimes very successful, but are less reliably so than littered nests. This is probably partly because of the lack of nesting material, despite the use in some designs of artificial grass, or 'astroturf', as a yielding substrate for hens to nest on. The sloping base and rolling away of eggs which are integral to these nests, however, are probably also aversive: blocking rollaways during early lay increases their use (Appleby, 1990). Nevertheless, nests with egg collection systems incorporating litter are potentially more appropriate for floor-housed poultry (Appleby *et al.*, 1988c). Two such systems are available. In 'Autonests', collection is achieved by moving eggs and litter together on a belt, then separating them and recycling the litter. In the other system, metal prongs rake through the litter, lifting the eggs onto a belt behind the nests.

The importance of loose nesting material for the welfare of poultry is controversial. While it is clear that hens prefer such material if it is available, the interpretation of this preference is difficult. Wild or feral birds do not nest in deep, loose material and it seems likely that, as with enclosure, hens are reacting to stimuli which are stronger than they would encounter in natural conditions. These are called supernormal stimuli (Appleby and McRae, 1986; Duncan and Kite, 1989). Nevertheless, provision of loose nesting material for hens is often recommended on welfare grounds.

A ratio of one nest for every four or five birds is usually recommended and a survey of broiler breeding houses found fewer floor eggs with this number than with higher ratios (Brocklehurst, 1975). In practice, ratios of from 1 : 6 to 1 : 8 are common with hens. In one extreme example, a broiler breeder flock was reported to have a nest ratio of 1:12.5, with access made even worse by dominance interactions between birds. This appeared to be the main cause of a severe floor laying problem in the flock, in which 30% of eggs were being laid on the floor (Perry *et al.*, 1971).

While all these aspects of management have indirect effects on nest-site selection, humans may also influence pre-laying behaviour directly, by training the birds. Most laying farms for hens carry out some training when the birds are first housed, in an attempt to prevent floor laying becoming established. Some methods could be called negative, such as disturbance of birds sitting on the floor and destruction of floor nests; it seems unlikely that these measures are effective. Positive methods, such as placing birds into nests, are more likely to work, especially if they can be confined there for a while: this is possible in trap nests and in some other designs with hinged perches. In one trial of this method, floor eggs laid by hens confined briefly in nests subsequently declined to 1%, compared to 24% in control pens (Craig, 1980). Even confinement for a period as short as 30 min greatly reduced floor laying in experimental flocks (Figure 11.6). However, it is probably rare for training in commercial flocks to be

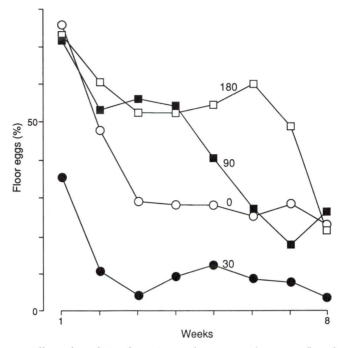

Figure 11.6.   Effect of confining hens in nest boxes on subsequent floor laying. In this Australian experiment, groups of broiler breeder hens were shut into nests for (○) 0, (●) 30, (■) 90 or (□) 180 min (Maguire, 1986).

carried out systematically. Reduced attention to individual birds in large flocks may explain the finding that floor laying is worse on average in pens of more than 900 birds than in smaller pens (Figure 11.3).

Nest-site selection is thus a complex process, but careful attention to all stages of the life history should allow it to be satisfactorily controlled. Once problems such as floor laying have become established, however, the conservatism of nesting birds often makes such problems difficult to cure.

11.6   Form of pre-laying behaviour

In housing systems with littered nest boxes, complete pre-laying behaviour similar to that in natural conditions is shown, with a searching phase, choice of a nest site and creation of a nest hollow (Wood-Gush, 1971). However, behavioural problems may still occur in such systems, depending on pen size and number of nest boxes. If nests are limited, aggressive interactions are common (Meijsser and Hughes, 1989) and floor laying may occur. By contrast, in large pens for hens, with many similar nest boxes, birds have difficulty in choosing between them and sometimes show pacing behaviour, which suggests that this

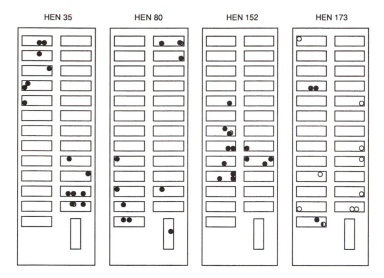

Figure 11.7.   Nest choice by commercial broiler breeders. Nesting behaviour of tagged hens was recorded on a number of days in a flock of nearly 4000 birds. Nest boxes were in blocks of 24, with 2 tiers of 6 on each side; they are shown here exaggerated in size. Open symbols = upper tier; filled symbols = lower tier (Appleby *et al.*, 1986).

difficulty is frustrating and deleterious to welfare (Appleby *et al.*, 1986). They also lay in different boxes on successive days (Figure 11.7). Difficulty in choosing between nest boxes probably also accounts for another feature of pre-laying behaviour which occurs in pens, namely gregariousness (Figure 11.8). Domestic hens and female turkeys will often enter occupied nests even if there are others free (Appleby *et al.*, 1984a). In extreme cases this can lead to breakage of eggs (which encourages egg eating) and even to birds being suffocated. It seems to arise very early in lay, in birds confronted with almost identical nest boxes. Occupied nests are the only ones which contrast with others and are investigated preferentially (Appleby and McRae, 1986). All these effects will be reduced in smaller pens, or in more heterogeneous houses which allow birds to localize their activities. Gregariousness is also avoided in turkeys by the use of semi-trap nests, with doors which prevent access by additional birds. These have not been used for hens, probably on grounds of cost.

In rollaway nest boxes, birds will usually settle as normal, but the behaviour associated with nest building is reduced in the absence of litter. Most rollaways are rectangular or oval in shape, but a round hollow has recently been suggested, to allow the rotating movements involved in nesting (Duncan and Kite, 1989). Some rollaway nest boxes, however, stimulate nest-site selection but not settled nesting: some birds will enter such boxes many times in a frantic manner before laying (Appleby, 1990).

Cages

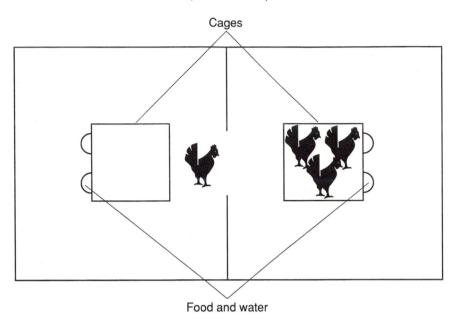

Food and water

Figure 11.8.   Gregarious nesting behaviour. In this experiment, hens were tested singly, with the choice of laying near flockmates or in isolation. The majority laid in the pen containing their flockmates (Appleby *et al.*, 1984a).

Such frantic pre-laying behaviour is, however, more characteristic of laying hens in conventional battery cages, where the searching phase is extended, in the form of restless pacing around the cage. Subsequent phases, though, differ between strains of hens (Wood-Gush, 1972). In most light hybrid strains, settled nesting is very brief or completely absent and hens continue to pace, often in a stereotyped way, until shortly before or even after oviposition. It is widely accepted that frustration of nesting is the most severe behavioural problem of hens in cages. It has recently been shown (Mills *et al.*, 1985a) that the extent of pre-laying pacing is under genetic control: its incidence can be reduced by appropriate selection. However, it is not known whether the lines which show decreased pacing are actually under less stress, or whether stress is simply not expressed in the same way. Medium hybrids, by contrast, usually sit before laying and often show 'vacuum' nesting behaviour: they go through the motions of making a nest hollow even though there is no litter. It has sometimes been suggested that this shows they are highly motivated to nest and that they are likely to be frustrated by absence of an appropriate substrate. However, this is not supported by records of heart rates, which suggest that they are calm during vacuum nesting (Mills *et al.*, 1985b). One other possibility is that they have little need of such a substrate (Appleby, 1990). At any rate, the calmer behaviour of medium hybrids compared to light hybrids in the pre-laying period is generally interpreted as better adaptation to the cage environment.

Frustration of nesting in cages can be avoided by the addition of nest boxes to modified designs (Robertson *et al.*, 1989; Appleby and Hughes, 1990), but the problems of automatic egg collection in such designs have not yet been wholly solved (Appleby, 1990).

Different strains of hens also show variation in other aspects of egg laying. In particular, floor laying is more common in medium hybrids than in light hybrids and in broiler breeders than in layer breeders. This is a reflection of the difference in mobility between these categories of bird and requires appropriate management (section 11.5).

11.7   Oviposition

Behaviour during egg laying itself may be an important cause of damage to eggs, particularly in cages. Individual birds vary in their laying stance, some habitually standing to lay, which tends to result in cracked eggs (Carter, 1971) although modern strains of hens have relatively short legs. In cages with perches, hens may lay from the perches, perhaps because these are more level than the sloping floor, or because they provide more support and this increases the problem (Duncan *et al.*, 1992). The problem is generally absent in systems, including modified cages, where eggs are laid in nest boxes containing litter or some other soft surface. An exception to this is in severe cases of gregarious laying when eggs are laid on top of other eggs.

In nest boxes, laying position can be a critical factor in the initiation of cannibalism (section 9.9). Vent pecking is more likely when birds face inwards to lay; this seems to be particularly common in rollaway nests which slope forwards, but it is not clear why birds should prefer to face up the slope rather than down.

If oviposition is delayed, retention of the egg in the shell gland often causes deposition of extra calcium on the surface. This gives a 'dusted' appearance, which is harmless to consumers but may nevertheless reduce the price paid by some buyers. It probably also reduces gaseous exchange through the shell and hence hatchability of fertile eggs. It sometimes occurs naturally, particularly early in lay, but is more frequent after disturbance during the pre-laying period. Disturbance to birds at an earlier stage also results in abnormal eggs, but in this case they are usually mis-shapen, presumably because of contractions in the shell gland before the shell has hardened (Figures 11.9, 11.10).

11.8   Post-laying behaviour

In cages or rollaway nest boxes, birds have little opportunity to sit on their eggs, although they sometimes do so for a while if eggs fail to roll away immediately. They also sometimes continue to pay attention to eggs even after they have

Figure 11.9. One of the possible effects of disturbance to birds while the shell of the egg is still soft; an equatorial bulge, caused by contraction of the shell gland.

rolled away. In littered boxes, by contrast, birds sit on eggs for a variable period, up to about half an hour. There is some indication that this results in poorer keeping quality for table eggs, by preventing them from cooling down so quickly. Sitting is also extended in rollaway nests, if false, 'decoy' eggs are fixed in them to encourage their use. In either of these circumstances, longer occupancy of nests means that providing the appropriate ratio of nest boxes to birds is even more important.

Allowing birds to sit on eggs also increases the likelihood of broodiness, even in highly selected laying strains. Appropriately, then, the usual treatment for broodiness in laying birds from a floor system, especially in turkeys, is to shut them in a cage or 'broody coop' on wire or slats. In most cases, they quickly resume laying. If broodiness is required, however, for example for incubation in small, farmyard flocks, it can be induced by introducing chicks to the females. In the past, the same effect has sometimes been achieved by depriving females of food and water for a period to induce moulting and interrupt laying. However, this is now regarded as unacceptable on welfare grounds (MAFF, 1987a).

One other problem which sometimes arises with littered nest boxes is egg eating. It is always initiated by accidental breakage of eggs, for example after floor laying or overcrowding of nest boxes, but once birds have experience of eating broken eggs they may learn to break more themselves. The problem is rarer in systems where eggs roll away, but not unknown: if eggs are cracked in battery cages, for example, hens can learn to peck them before they roll out of reach. As with other behavioural problems, this is difficult to cure once it has become established, so it is important to prevent it by good management. Reduction of light levels, however, may help to reduce the problem. Conversely, bright light seems to be a contributing factor, perhaps through a general effect on activity.

11.9   Egg production

Although food intake and food conversion efficiency vary between systems (Chapter 8), no clear differences in egg production have been found. For laying hens, however, a difference which does exist is loss of eggs in more extensive systems, from causes such as floor laying. As a result, there has been a slight tendency for more eggs to be collected in cages than in other systems. This has not been consistent (Table 11.1) and should be amenable to management.

Within systems, it is common for egg production to decline with increased

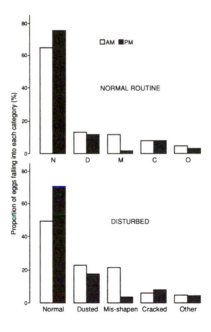

Figure 11.10.   The effect of disturbance to birds on their eggs is affected by the stage of egg development. The incidence of shell abnormalities from hens housed on deep litter was recorded on days following a normal routine (upper diagram) and following a major disturbance (lower diagram). Most eggs collected in the morning (open columns) would have been in the shell gland during the disturbance and a high proportion were mis-shapen. Eggs collected in the afternoon (filled columns) would have been ovulated later (Hughes *et al.*, 1986).

stocking density and group size. For laying hens, this has been most clearly demonstrated in cages (Hughes, 1975b); it has also been recorded in a straw-yard, but not in deep litter (Table 11.1). Depression of egg production has sometimes been interpreted as an indicator of poor welfare; this is controversial (Appleby and Hughes, 1991), but in the studies of cages cited by Hughes (1975b) egg production correlated with more reliable indices such as mortality.

Egg production is unique in animal production systems. The product, whether used for human consumption or for breeding, is conveniently packaged and has an integral delivery mechanism in the form of 'pre-programmed' egg laying behaviour. This behaviour is also unusual in the extent to which we understand it, so it is ironic that control of egg laying was one of the major factors in the development of battery cages. Such understanding can now be put to good use in management of other systems.

Table 11.1. Egg numbers from ISA Brown hens in alternative systems and from comparable birds in cages. It should be noted that results from different studies are not directly comparable and that not all of these studies compared systems statistically.

System	Ages (weeks)	Stocking density (Birds/m²)	Year 1	Year 2	Year 3	Reference
Free range	20–68	0.1 (3 in house)	245			Hughes and Dun, 1986
	20–72			283	287	
Cages	20–68	21	251			
	20–72			280	284	
Strawyard	20–72	3	261			Gibson *et al.*, 1988
		4	247	288	291	
		5		283		
		6			285	
Deep litter	20–64	3	224			Appleby *et al.*, 1988b
		6	228			
		7		208		
		8		224		
		10		235	225	
		11			232	
Cages	20–64	13	242	234	253	
		18	236	230	248	
Perchery	20–44	13	133			McLean *et al.*, 1986
Cages		18	137			

12

Movement and maintenance

12.1 Summary

- Movement and maintenance behaviour is relevant to welfare, in that it is involved in maintaining the physical and mental health of the birds. Both domestic and waterfowl are active species under natural conditions, moving considerable distances each day while feeding and foraging.
- Cages constrain the freedom to exercise most normal patterns of behaviour and at high stocking densities do not allow animals to turn around and stretch their limbs without difficulty. Walking is infrequent in cages compared to alternative systems and wing flapping non-existent. As a probable consequence, bone strength in cages is less.
- In floor pens, even when groups are large, birds tend not to have home ranges, but rather to move throughout the flock. In large groups they will thus be constantly encountering unfamiliar individuals.
- Comfort behaviour, such as preening, stretching and feather ruffling, is constrained at high stocking densities and also by low cage heights. It shows a rebound in frequency when birds are moved to more spacious conditions. Dust bathing normally serves to maintain plumage in good condition; it occurs in 'vacuum' form in cages. Because it is diurnal in frequency, in some systems dust baths are opened only after mid-day, to prevent laying in them during the morning.
- Scratching and pecking are common behaviour patterns; the former has been utilized, in conjunction with abrasive strips, to control claw length in cages. Pecking in barren surroundings may be directed towards the plumage of other birds. Fowls rest and roost on perches, where available; perches can have beneficial effects on foot condition, if fitted to cages.

● Hens in cages tend to be more fearful than those in more varied alternative systems. Excessive fearfulness and over-reaction to unusual stimuli can be a problem, especially in large groups kept in barren conditions, such as colony cages or wire floored enclosures.

12.2   Natural behaviour

Wild and feral poultry are active birds. For example, red jungle fowl in semi-natural conditions spend the majority of the day foraging even if they are supplied with adequate food (Dawkins, 1989). The jungle fowls and the other galliforms such as pheasants and feral hens use a fairly well defined home range (Collias and Collias, 1967; Wood-Gush *et al.*, 1978). They become familiar with this area and know the best places for feeding and resting. Roosting at night occurs in regularly used bushes or trees and shorter rests in the day are usually also off the ground but in more variable locations. Flight is infrequent except in going up to such resting places or descending and, although chicks are often very mobile, adults usually walk unless running or flying is necessary. Waterfowl are similarly active in foraging, either in water or on land and are generally more ready to use flight for moving around. As a result, they often use larger areas from day to day.

In the variable environmental conditions encountered, it is important for birds to keep their plumage in good condition and this is achieved by frequent preening and other comfort behaviour. In the waterfowl this behaviour also acts to waterproof the feathers. Wild birds nearly always look well groomed and wild and feral poultry are no exception in this respect.

12.3   Behavioural expression

In discussions of the welfare of animals in intensive husbandry systems, one of the most frequent considerations is restriction of behaviour. This is particularly true in criticisms of conventional cages for laying hens. Cages apparently contravene one of the UK Farm Animal Welfare Council's five freedoms (see Preface), namely the freedom to exercise most normal patterns of behaviour. Furthermore, it is generally accepted that they contravene the more basic principle of the UK's Brambell Committee (section 6.3) that 'An animal should at least have sufficient freedom of movement to be able without difficulty to turn round, groom itself, get up, lie down and stretch its limbs' (HMSO, 1965). In comparison, observations in alternative systems for laying hens show that they engage in a wide variety of behaviour patterns (Figure 12.1; Gibson *et al.*, 1988). Of course, to assess the importance of behavioural expression for welfare, it is necessary to have an understanding of which behaviour patterns are important to the birds. There has been considerable progress towards such an

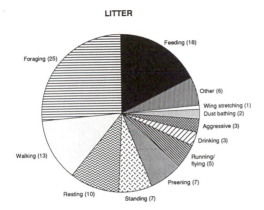

LITTER

Feeding (18)
Foraging (25)
Other (6)
Wing stretching (1)
Dust bathing (2)
Aggressive (3)
Drinking (3)
Walking (13)
Running/ flying (5)
Preening (7)
Resting (10)
Standing (7)

Figure 12.1.   Behaviour of laying hens in a deep litter house: proportion of time spent in different activities on the litter (top) and on a raised, slatted area (bottom) (Appleby *et al.*, 1989).

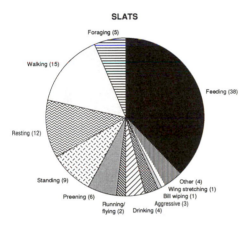

SLATS

Foraging (5)
Walking (15)
Feeding (38)
Resting (12)
Standing (9)
Other (4)
Wing stretching (1)
Bill wiping (1)
Preening (6)
Aggressive (3)
Running/ flying (2)
Drinking (4)

understanding (Dawkins, 1983, 1988; Webster and Nicol, 1988; Appleby, 1990) and certain behaviour patterns are now most often the focus of attention. These include pre-laying behaviour (Chapter 11) and others considered in the following sections.

12.4   Movement

The word 'movement' and associated consideration of freedom of movement, combines two aspects of behaviour which can usefully be thought of separately: the relatively small-scale actions involved in actual performance of all behaviour, and larger-scale locomotion. The Brambell Committee's use of the phrase 'freedom of movement', quoted above, is in terms of small-scale actions. Even in these terms, measurement of the area occupied by hens has shown that conventional battery cages must restrict freedom of movement. A medium hybrid hen, unconstrained and including the tail and other feathers, occupies between

475 and 600 cm^2 at rest and more if active (Bogner *et al.*, 1979; Freeman, 1983; Dawkins and Hardie, 1989). This area will be affected by posture and part of it may protrude beyond nominal space allowances, but at the minimum EC allowance of 450 cm^2 hens must frequently overlap or have their feathers compressed. Crowding will restrict behaviour; it may also be directly detrimental to welfare (Mashaly *et al.*, 1984). No other system, for any poultry, is so restrictive as battery cages. For laying hens, provision of even 600 cm^2 per bird would allow 16.7 birds per m^2 and in no single tier floor system is it recommended that birds are stocked as densely as this.

Freedom of movement is reflected in the actual number of movements made by birds. One study comparing different systems (Knowles and Broom, 1990) found that hens took an average of 72 steps per h in cages, 208 in a perchery and 1058 in an Elson Terrace. Wing movements were, however, most common in the perchery, with wing flapping twice per h and flying 0.4 times per h, and almost completely absent in the other systems. Another study found similar differences in wing flapping between hens in deep litter or Hans Kier systems and those in cages (Nørgaard-Nielsen, 1990). These differences affect bone strengths. Tibia strength is increased by up to 41% and humerus strength by up to 85% in percheries, deep litter, Elson Terrace and Hans Kier systems, compared to cages (McLean *et al.*, 1986; Knowles and Broom, 1990; Nørgaard-Nielsen, 1990). Bone strength and structure may also be improved in cages, simply by adding a perch, although probably not as much as in alternative systems (Figure 12.2; Hughes *et al.*, 1992). Bone strength is an important feature for both welfare and economics, because weak bones are more likely to be broken both within the system and when birds are removed for slaughter. Up to 30% of caged birds suffer broken bones during catching and transportation and more during processing (Gregory and Wilkins, 1989). There are around half as many breakages in birds from free range or percheries (Gregory *et al.*, 1990).

Restriction of movement will also result in the prevention of specific behaviour patterns, because these need more space than standing (Figure 12.3; Table 12.1). Such prevention may cause frustration, as discussed later in the chapter.

The question of whether larger-scale locomotion is also important to birds prompts contrasting answers. Members of the general public tend to believe that poultry and other animals should be able to move widely. In contrast, there is a widespread opinion among poultry farmers that birds in deep litter or similar systems will not move long distances even to reach facilities such as feeders or nest boxes and that these must therefore be distributed evenly through the house. In fact, there is little evidence for either of these beliefs. Broiler breeder chickens will move readily for nesting or other behaviour (Figures 9.4, 11.7) and even broilers, which are commonly regarded as almost immobile in commercial conditions either by nature or because of crowding, actually move about extensively (Figure 12.4). However, there is no evidence that such movements are necessary to birds which have adequate facilities close by. There have been numerous studies of the behaviour of hens in get-away cages and other modified

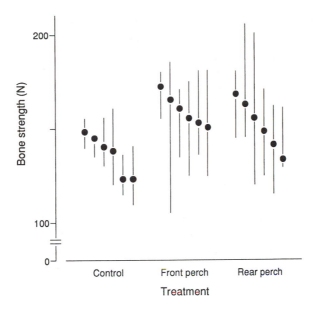

Figure 12.2. Tibia strength in Newtons (N) of hens at end of lay, from cages with no perches, front perches or rear perches, measured by breaking the bones on a three point rig. There were six cages in each treatment with four birds per cage; medians and ranges of values from each cage are shown (Hughes and Appleby, 1989).

cage designs (Appleby and Hughes, 1990; Wegner, 1990), with no indication of frustration from restriction of movement. It does not seem that poultry have any particular motivation for exploration except in connection with foraging for food or searching for other resources such as nest sites.

It remains true, however, that restriction of large-scale movement is often associated with restriction of small-scale movement, because of the confounding effect of stocking density. For example, in one study comparing cages with a perchery, hens were active for a similar proportion of time (0.91 and 0.85 respectively) but the mean distance moved was seven times greater in the perchery (McLean *et al.*, 1986). Mobility is directly affected by density: in a deep litter system studied over a range of stocking densities, time spent in locomotion declined at higher densities (Figure 12.5). However, not all movement is beneficial. When hens from one perchery were dissected, some were found to have bones which had been broken, then healed (Gregory *et al.*, 1990). This may have been because they had to jump a large gap to reach the nest boxes and some failed to do so successfully (Broom, 1990). Clearly, future designs of percheries and aviaries should take this problem into account.

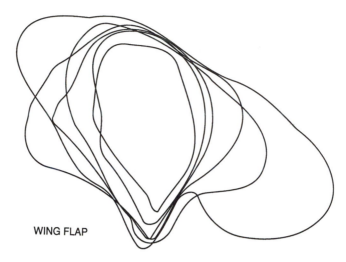

Figure 12.3.   The space used for wing-flapping by an unrestricted hen. Successive outlines of birds were drawn from an overhead video picture, starting with the smallest outline when the bird was standing still (Dawkins and Nicol, 1989).

12.5   Use of space

The belief among producers that poultry in large houses will not move far probably stems from the idea that they use well-defined home ranges, similar to those in the wild but much smaller. This was put forward by McBride and Foenander (1962), who suggested that birds in large flocks would restrict their movements to small areas in which they could recognize other individuals. In fact, home ranges are usually either ill-defined or maintained by only a few birds. For example, in both deep litter and strawyards, pen area is used unevenly

Table 12.1. Area used by medium hybrid hens housed singly in small litter-floored pens (from Dawkins and Hardie, 1989).

	Area (cm^2)	
Behaviour	Mean	Range
Standing	475	428–592
Ground scratching	856	655–1217
Turning	1272	978–1626
Wing stretching	893	660–1476
Wing flapping	1876	1085–2606
Feather ruffling	873	609–1362
Preening	1151	800–1977

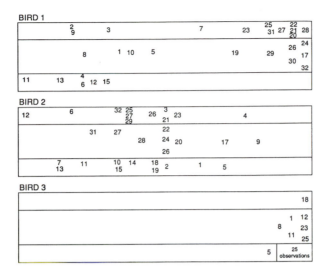

Figure 12.4.   Movements by individual broilers in a flock of over 18 000. Numbers show successive, twice-daily locations from day 27 to day 42. Some moved gradually along the house (bird 1), some moved up and down frequently (bird 2) and some moved very little (bird 3) (Preston and Murphy, 1989).

by individuals and by the whole flock, but individual ranges overlap extensively (Gibson *et al.*, 1988; Appleby *et al.*, 1989). In a commercial deep litter house for 4000 broiler breeder chickens, individual ranges averaged 73% of the area (Figure 9.4). This will mean that in most flocks too large for flockmates to know each other, birds are constantly surrounded by unfamiliar individuals. This does

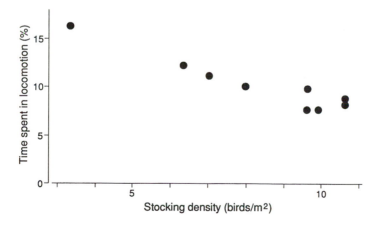

Figure 12.5.   Effect of stocking density on mobility in nine deep litter flocks (Appleby *et al.*, 1989).

not generally seem to cause aggression, but other possible effects and possible remedies are discussed in section 9.6.

In small pens and in cages, use of space is affected by social rank. Among laying chickens in pens, both males and females of high rank have smaller ranges than those of low rank (van Enckevort, 1965; Pamment *et al.*, 1983), presumably because the latter are avoiding the former while reaching facilities such as feeders. Similarly, in the cages studied by Keeling and Duncan (1989) the top-ranking hen was able to use preferred areas for a greater proportion of the time than other birds. This effect means that constraints on behavioural synchrony (section 9.4) are worse for low-ranking birds.

12.6   Comfort behaviour

Preening and other comfort behaviour such as wing flapping, feather ruffling and stretching, is important in artificial conditions for keeping the plumage well groomed, just as in natural conditions. This behaviour varies between systems in frequency, in form, in synchrony and to some extent also in function. This variation is primarily associated with stocking density, because comfort behaviour requires a large area for performance (Figure 12.3; Table 12.1). In hens it is therefore less frequent in battery cages than in more spacious systems and less frequent in small cages than large ones (Nicol, 1987a, b). To a lesser extent it is also constrained by cage height (Nicol, 1987a) and in fact the cage height of 35 to 40 cm required by the EC restricts quite a lot of behaviour. With unrestricted height, nearly 25% of hens' head movements occur above 40 cm (Figure 12.6). When hens are moved out of small cages, they perform comfort behaviour at an increased frequency, which suggests that these constraints cause frustration (Nicol, 1987b).

Some comfort behaviour varies in form with space allowance. In particular, preening can be performed in less space at high stocking density than when unrestricted (Dawkins and Hardie, 1989). However, it is likely to be less efficient, especially as close contact between birds and against the wire will frequently result in feathers being out of place. This may exacerbate the problem of feather pecking which is on average worse in cages than in other systems (although variable in the latter; section 9.8). Despite the use of less space, preening is also less synchronous in small cages than in large ones, with birds often preening on their own rather than all together (Jenner and Appleby, 1991).

Comfort behaviour has an additional relevance to welfare because it does not always seem to be functional: or rather, it appears in contexts which suggests it has functions in addition to increasing body comfort. Thus birds which are prevented from reaching food often preen themselves, but in a slightly faster, more incomplete manner than normal (Duncan, 1970). Similarly, penned hens which are accustomed to nest boxes show an increased frequency of wing flapping if those boxes are closed (Duncan, 1982). In these contexts, such performances of

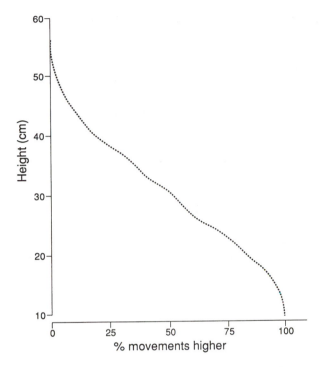

Figure 12.6. The height of head movements by hens in cages of unrestricted height. Birds were filmed from the side, and each time a bird's head moved, the height it moved to was recorded (Dawkins, 1985).

behaviour, apparently in an inappropriate context, are called displacement activities and are interpreted as indicating frustration. Despite space restriction, they are seen more commonly in highly intensive systems than in others.

12.7   Dust bathing

The comfort behaviour of dust bathing is different from others in that it involves dust or other loose material in its complete form. Fluttering movements work this material up into the feathers, where it helps to distribute or remove oily secretions and may help to control parasites. Dust bathing therefore occurs most often in housing systems with loose material. However, it can also occur in a 'vacuum' form, in which the bird carries out the same actions on slats or wire, although more briefly (Vestergaard, 1980). This is sometimes interpreted as indicating high motivation, in which case birds deprived of loose material might suffer frustration. Extensive experiments, though, have failed to find any other evidence for such strong motivation (Dawkins and Beardsley, 1986). In contrast, an alternative explanation for vacuum behaviour is that it may indicate a

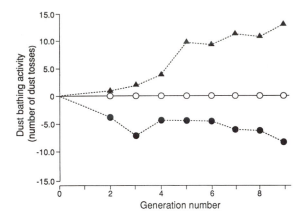

Figure 12.7.   Selection for high (▲) and low (●) dust bathing activity in quail. Average values for behaviour during observations are expressed as deviations from that of birds of a control line (○) (Gerken and Petersen, 1987).

low behavioural threshold. For example, it is probable that dust bathing can be triggered simply by the sight of food, which is certainly a loose material. It seems likely, in fact, that this behaviour is partially 'stimulus bound': increased by the presence of an appropriate stimulus but forgotten in its absence – 'out of sight, out of mind'. To the extent that this is true, birds will not suffer behaviourally from the absence of loose material for dust bathing. It is not completely true, though, because dust bathing is also affected by the condition of the plumage: it is increased by the presence of fatty material on the feathers (van Rooijen, 1989). Lack of loose material in some systems is often assessed as a welfare problem (Vestergaard, 1982; Farm Animal Welfare Council, 1986). This may be justified partly because dust bathing has physical as well as behavioural effects and partly because loose material is used for other behaviour, as in the next section.

Dust bathing has a strongly diurnal rhythm, being mostly confined to the afternoon. This has the result that in partly littered systems the litter may be deserted in the morning but crowded later in the day. In one perchery for laying hens, birds clustered at 24 per m² for this purpose (McLean *et al.*, 1986). This rhythm is used in the Elson Terrace and in some designs of modified cages, by allowing hens access to loose material only in the afternoon. Rollaway nest boxes can then be provided without risk of hens laying in the dust bathing areas, because most laying occurs in the morning. In quail, however, laying also occurs in the afternoon so this option is not available. Selection experiments in quail have shown that dust bathing is under close genetic control. Lines were produced which showed increased or decreased dust bathing compared to controls (Figure 12.7). Given the equivocal nature of dust bathing motivation discussed above, however, it seems unlikely that such selection would be worthwhile commercially, even for welfare reasons.

Birds readily dust bathe in wood shavings or other floor litter, but if finer material such as sand is available they use this in preference. In an experimental study of modified cages for laying hens, dust baths were filled with sand and nest boxes with wood shavings. No dust bathing was recorded in the nest boxes and few eggs were laid in the dust baths (Robertson *et al.*, 1989).

12.8   Pecking and scratching

Pecking and scratching also generally use loose material, if it is available. These are common activities of poultry, forming an important part of the foraging behaviour discussed above (section 12.2). In the most common form, chickens and other galliforms scratch with both feet then move quickly backwards, pecking at anything edible exposed by the scratching. In a deep litter house, hens spent 25% of their time on litter in this behaviour in addition to time at the feeders (Figure 12.1). It is not clear whether birds can be trained to perform various tasks to earn access to loose material, in order to peck and scratch. However, motivation for such behaviour is apparently not stimulus bound, as dust bathing seems to be. Parts of it are an integral part of feeding behaviour, even where they seem to be inappropriate: for example, chickens scratch while feeding, even if feeding on freely available food in a metal trough. This tendency has actually been used to advantage in battery cages, in a particularly good example of how management can utilize behaviour. If a strip of abrasive tape is fixed to the manure deflector behind the food trough, scratching by the hens prevents overgrowth of their claws and the damage and injury which this can produce (Tauson, 1986). This adaptation to cages is now compulsory in Sweden.

The absence of varied or manipulable substrates in systems such as cages, however, leads to other problems. Perhaps most importantly, it probably contributes to the incidence of feather pecking and cannibalism in cages: in one experiment, pullets in pens which were deprived of such substrates showed an increased frequency of redirected pecking (Blokhuis, 1989). It may be possible to provide stimuli in cages, such as coloured spots, which increase ground pecking and reduce feather pecking (Braastad, 1990). However, it will probably be more effective to provide loose material. Even in cages provision of a dust bath allows dust bathing, pecking and scratching and results in better feather and foot condition (Robertson *et al.*, 1989).

If there is edible matter in the substrate, pecking and scratching are considerably increased. It used to be common to scatter some grain in deep litter houses and strawyards for this reason, to help keep the litter in good condition. Poultry also eat indigestible matter such as sand or fine grit. Some of this is retained in the gizzard and, in free range birds and others which are fed large food items, helps to grind the food.

12.9   Rest and sleep

The main pattern of rest and sleep is set by the light–dark cycle. As with most birds, poultry are generally inactive at night and this diurnal rhythm is strengthened in enclosed houses with a completely dark night, compared to systems with natural lighting and more gradual dawn and dusk. It is weakened, by contrast, in light regimes which use a dim phase rather than complete darkness and it is disrupted even more by intermittent lighting programmes without a well-defined night (Blokhuis, 1983; Coenen *et al.*, 1988). There has been very little consideration of the effects on the birds of such disruption, but no definite ill-effects are known. In a normal day–night cycle, other periods of rest occur at intervals through the day, usually with some synchrony between neighbours. This can be readily understood by reference to a farmyard flock or other free-ranging group, in which coordinated behaviour is necessary if they are to stay together.

If perches or roosts are available, chickens and other galliforms usually use them for night-time roosting and also for resting in the day if they are easily accessible (Figure 12.8). Among hens kept on wire, this improves the condition of their feet (Figure 12.9) and fitting a perch in cages also tends to benefit feather condition (Duncan *et al.*, 1992). In floor-housed systems, use of perches is affected by experience in at least some strains of hens. If these birds are reared with perches, however, they readily learn to perch and this may have an incidental effect in reducing later floor laying (section 11.5).

When perch space is limited, the struggling of hens to get on to perches at dusk is often vigorous, suggesting that this is a strongly motivated behaviour pattern. Perhaps because of this strong motivation, resting birds frequently crowd very closely on to perches, particularly when they are young and not yet of full body size. For all hens to perch once they are full grown, however, an allowance of about 150 mm of perch per bird is necessary for most strains, possibly less for light hybrids. If more space is available they will, in fact, space more widely than this. There have been some attempts to design housing systems with mobile barriers, to take advantage of different day and night space requirements, but these have so far proved impractical. In the Dwelltime system, for example, two flocks alternated between a small, dark pen and a large, light pen.

12.10   Fearfulness

Poultry react to novel, potentially threatening stimuli in ways which depend both on the stimuli and on other features of the environment, especially the housing system. A common response to mildly threatening stimuli, such as the appearance of unfamiliar humans, is headshaking (also sometimes called headflicking). This seems to indicate alerting, or attention (Hughes, 1983b).

lation that poultry welfare must be taken into account in choosing and managing production systems.

In terms of welfare, all systems offer potential advantages and disadvantages; the advantages are not always achieved and the disadvantages are not always minimized. Nevertheless, it does seem that there are more welfare problems in some systems than in others. This can be illustrated by considering systems in relation to the five 'freedoms' listed by the UK's Farm Animal Welfare Council (see Preface). All systems can provide 'freedom from hunger and thirst', or at least minimize those conditions. Many systems, however, including extensive systems, have problems associated with two of the other four areas: 'freedom from pain, injury and disease' and 'freedom from fear and distress'. Indeed, these problems may be worse on occasion in extensive systems than in more intensive systems, for example because of aggression and cannibalism associated with large group size, or because of the preventative beak trimming which is itself a major welfare problem. When these problems occur in extensive systems they are often more difficult to control; because such a system by its very nature is not a controlled environment, producers do not have recourse to strategies such as reducing light intensities, or subdividing flocks.

However, highly intensive systems, such as conventional cages for laying hens, are more likely to compromise not only the previous two freedoms, but also the remaining two: 'freedom from discomfort' and 'freedom to exercise normal behaviour'. Ideal systems should combine the small group sizes found in some intensive systems with the freedom of movement and complex environment of more extensive systems. Intermediate systems, such as modified cages or other small units for laying hens and small pen sizes for many other types of poultry, justify more research and development than they have so far been accorded.

An increasing consideration in matching poultry biology and environment is genetic selection. Many studies have been mentioned in the appropriate chapters on whether advantageous traits can be increased, or disadvantageous traits decreased, by selection. These studies form part of a wide debate on whether poultry can be adapted genetically to suit commercial environments, or whether the main emphasis should be on providing appropriate environments for the birds as they are (Faure, 1980). There are two main difficulties with the former approach. First, new developments in housing and husbandry of poultry tend to occur too often for behavioural selection to keep up with them. A major selection programme would be necessary to alter behaviour for a particular production system, while controlling for production traits, and by the time it had achieved significant changes it would probably be redundant. Second, for most poultry the environment in which selection occurs is not the same as that used for actual production. Thus in chickens, housing conditions for breeding flocks are very different from those for broilers or layers. For this reason, the common idea that farm animals must have become progressively more adapted to production environments over many generations is mistaken and such adaptation is unlikely to be a significant effect. This is demonstrated by the fact that some breeds of hens can survive and reproduce in feral conditions (Duncan *et al.*, 1978).

Selection is costly, but where behaviour affects production it may be economic to select for it in future. In particular, genetic selection against behavioural traits which can affect both production and welfare and which, in addition, can cause problems in cages as well as in more extensive systems, would be worthwhile. Examples of such traits include propensity to feather peck or cannibalize, to direct aggression against other birds, or to show high levels of fear or in extreme cases, hysteria. Selection for behaviour which affects welfare but not production, on the other hand, is unlikely to happen without legislation or assistance from public funds. One area in which change has already occurred, however, is in choice of particular strains of poultry which differ in behaviour. In the UK and in parts of the USA, change from production of white hens' eggs to brown occurred as a result of customer preference. A side effect was that the behaviour of laying hens in these countries changed; this was because strains of hens which produce brown eggs are more placid than white-egg layers. However, some brown hybrids show a stronger propensity for feather pecking and cannibalism than white-egg strains. Choice of strains appropriate to different systems has potential which could be further investigated.

The behaviour of birds can also, of course, be modified within their lifetime, by appropriate management and husbandry. Here attitudes are important, as well as the knowledge and skills referred to throughout the book. This is particularly clear in relation to the sorts of abnormal behaviour which are sometimes referred to as 'vices'. As pointed out in Chapter 9, use of this term suggests that the birds are themselves 'to blame' and that the problems are likely to be intractable. In fact, increased understanding of such behaviour can help to prevent it occurring and reduce its effects if it does occur. In particular, the conditions under which birds are reared may be very important. Increased education and training of people responsible for looking after poultry, in whichever system they are kept, will have a major beneficial influence on their welfare. Further, attitudes of those in charge of management and husbandry will be major determinants of both welfare and profitability.

Choice and management of less intensive production systems, selection of stock and training of personnel all involve increased expenditure. In countries where specialist markets can be developed, increased cost may in some circumstances be offset by increased income from specialist products, such as free range eggs and broilers. As emphasized in Chapter 5, however, these markets are limited. For the rest of the poultry industry, the details of what happens in future must therefore depend on the interactions between economics and legislation. We have suggested that the cost of a system and the welfare of birds kept in it will be approximately proportional (Appleby and Hughes, 1991). It follows that, within a system, competition between producers on price, resulting in attempts to reduce costs by increasing the number of birds in a house or by reducing input of labour per bird is likely to decrease welfare. If such measures are restricted by legislation, costs will be incurred. It is apparent that such costs must be borne by one or more of three sources: producers, purchasers or public funds. Given the public consensus in most countries that such legislation is appropriate, we believe that the balance between these sources should be considered carefully by the producers, by the public and by the governments which control the use of taxpayers' money.

It is generally assumed that effects of legislation relevant to animal welfare will be divided between producers and purchasers. However, the ways in which this can be achieved are limited and will have far-reaching significance. In particular, reduction of profit margins available to producers would have implications for the size of viable production units, itself a matter for debate. On the other hand, increased price of poultry products in the shops would affect sales (although less so for eggs than for some other products) and the standard of living of people who regard these as staple foods. The third possibility, that the costs of legislation on animal housing should be met partly by subsidy from public funds, has been less considered than other options. It could be argued that if society is concerned with the conditions in which animals are kept, society should be willing to contribute to the costs of their upkeep. Use of public funds in this way would spread those costs more widely than insisting that producers and purchasers should pay the whole cost. It could also usefully be associated with existing or expanded programmes of farm inspection, so that such inspection could result in reward rather than just punishment as at present. In any event, it is clear that informed debate on such matters is necessary before decisions are made. There has been an increase in such debate among those most actively interested, but we suggest that it needs to be broadened still wider to include elements of public education and assessment of public opinion.

The subject matter of this book and of these Conclusions has been restricted to poultry, but it is evident that many or most of the general principles and arguments apply equally to the production and welfare of other animals. Indeed, although more research and development have been carried out on poultry than on other farm animals, there is sufficient information available to take a similar approach in all animal production systems. This emphasizes the importance of both strategic and applied research in supporting the animal production industry, an importance which can only increase in future.

References

Agriculture Canada (1989) *Recommended Code of Practice for the Care and Handling of Poultry from Hatchery to Processing Plant.* Publication 1757E. Ottawa, Agriculture Canada

Alvey, D.M. (1989) Evaluation of two perchery designs for egg production. *British Poultry Science,* **30**, 960

American Poultry Association (1989) *American Standard of Perfection.* American Poultry Association, Inc., Estacada, Oregon

Amgarten, M. and Mettler, A. (1989) Economical consequences of the introduction of alternative housing systems for laying hens in Switzerland. In *Proceedings, Third European Symposium on Poultry Welfare* (Faure, J.M. and Mills, A.D. eds), pp. 213–228. Tours, WPSA

Anderson, D.P., Beard, C.W. and Hanson, R.P. (1964) Adverse effects of ammonia on the surface ultrastructure of the lung and trachea of broiler chickens. *Poultry Science,* **64**, 2056–2061

Anon. (1983) Perchery tries again to match cages. *Poultry World* 12th August: 10

Appleby, M.C. (1984) Factors affecting floor laying by domestic hens: a review. *World's Poultry Science Journal,* **40**, 241–249

Appleby, M.C. (1985) Hawks, doves . . . and chickens. *New Scientist* **1438**, 16–18

Appleby, M.C. (1990) Behaviour of laying hens in cages with nest sites. *British Poultry Science,* **31**, 71–80

Appleby, M.C. and Duncan, I.J.H. (1989) Development of perching in hens. *Biology of Behaviour,* **14**, 157–168

Appleby, M.C. and Hughes, B.O. (1990) Cages modified with perches and nest sites for the improvement of welfare. *World's Poultry Science Journal,* **46**, 38–40

Appleby, M.C. and Hughes, B.O. (1991) Welfare of laying hens in cages and alternative systems: environmental, physical and behavioural aspects. *World's Poultry Science Journal,* **47**, 109–126

Appleby, M.C. and McRae, H.E. (1986) The individual nest box as a superstimulus for domestic hens. *Applied Animal Behaviour Science,* **15**, 169–176

210

Appleby, M.C., Duncan, I.J.H. and McRae, H.E. (1988a) Perching and floor laying by domestic hens: experimental results and their commercial application. *British Poultry Science*, **29**, 351–357

Appleby, M.C., Hogarth, G.S., Anderson, J.A., Hughes, B.O. and Whittemore, C.T. (1988b) Performance of a deep litter system for egg production. *British Poultry Science*, **29**, 735–751

Appleby, M.C., Hogarth, G.S. and Hughes, B.O. (1988c) Nest box design and nesting material in a deep litter house for laying hens. *British Poultry Science*, **29**, 215–222

Appleby, M.C., Hughes, B.O. and Hogarth, G.S. (1989) Behaviour of laying hens in a deep litter house. *British Poultry Science*, **30**, 545–553

Appleby, M.C., McRae, H.E. and Duncan, I.J.H. (1983) Nesting and floor laying by domestic hens: effects of individual variation in perching behaviour. *Behaviour Analysis Letters*, **3**, 345–352

Appleby, M.C., McRae, H.E., Duncan, I.J.H. and Bisazza, A. (1984a) Choice of social conditions by laying hens. *British Poultry Science*, **25**, 111–117

Appleby, M.C., McRae, H.E. and Pietz, B.E. (1984b) The effect of light on the choice of nests by domestic hens. *Applied Animal Ethology*, **11**, 249–254

Appleby, M.C., Maguire, S.N. and McRae, H.E. (1985) Movement by domestic fowl in commercial flocks. *Poultry Science*, **64**, 48–50

Appleby, M.C., Maguire, S.N. and McRae, H.E. (1986) Nesting and floor laying by domestic hens in a commercial flock. *British Poultry Science*, **27**, 75–82

Aschoff, J. and Meyer-Lohmann, J. (1954) Angeborene 24-Stunden-Periodik beim Kuken. *Pflugers Archiv fur die gesamte Physiologie des Menschen und der Tiere*, **260**, 170–176

AWI (Animal Welfare Institute) (1991) *Animals and Their Legal Rights. A Survey of American Laws from 1641 to 1991*. Washington, D.C., Animal Welfare Institute

Baker, K.B. (1988) Legislation now and for the future. In *Cages for the Future*, pp. 1–10, Cambridge Poultry Conference, Agricultural Development and Advisory Service

Bang, B.G. and Wenzel, B.M. (1985) Nasal cavity and olfactory system. In *Form and Function in Birds* (King, A.S. and McLelland, J. eds) Vol. 3, pp. 195–225. London, Academic Press

Baxter, M. R. (1983) Ethology in environmental design. *Applied Animal Ethology*, **9**, 207–220

BEIS (British Egg Information Service) (1990) *Egg Farming*. London, British Egg Information Service

Bell, D.J. and Freeman, B.M. (1971) *Physiology and Biochemistry of the Domestic Fowl*, Vols 1–3. London, Academic Press

Belshaw, R.H.H. (1985) *Guinea Fowl of the World*. Liss, Hampshire, Nimrod Book Services

Bessei, W. (1973) Die selective Futteraufnahme beim Huhn. *Deutsche Geflugel-wirtschaft und Schweineproduktion*, **25**, 107–109

Bessei, W. (1986) Das Verhalten des Huhns in der Intensivhaltung. *Jahrbuch der Geflugel produktion*, 95–99

Bhagwat, A.L. and Craig, J.V. (1975) Fertility from natural matings influenced by social and physical environments in multiple-bird cages. *Poultry Science*, **54**, 222–227

Biggs, P.M. (1990) Vaccines and vaccination – past, present and future. *British Poultry Science*, **31**, 3–22

Blokhuis, H.J. (1983) The relevance of sleep in poultry. *World's Poultry Science Journal*, **39**, 33–37

Blokhuis, H.J. (1984) Rest in poultry. *Applied Animal Behaviour Science*, **12**, 289–303

Blokhuis, H.J. (1989) The effect of a sudden change in floor type on pecking behaviour in chicks. *Applied Animal Behaviour Science*, **22**, 65–73

Blount, W.P. (1951) *Hen Batteries*. London, Baillière Tindall and Cox

Bogner, H., Peschke, W., Seda, V. and Popp, K. (1979) Studie zum Flaschenbedarf von Legehennen in Kafigen bei Bestimmten Activitaten. *Berliner und Munc23her Tierarztliche Wochenschrift*, **92**, 340–343

Braastad, B.O. (1990) Effects on behaviour and plumage of a key-stimuli floor and a perch in triple cages for laying hens. *Applied Animal Behaviour Science*, **27**, 127–139

Braastad, B.O. and Katle, J. (1989) Behavioural differences between laying hen populations selected for high and low efficiency of food utilisation. *British Poultry Science*, **30**, 533–544

Brake, J. (1987) Influence of presence of perches during rearing on incidence of floor laying in broiler breeders. *Poultry Science*, **66**, 1587–1589

Bremond, J.C. (1963) Acoustic behavior of birds. In *Acoustic Behavior of Animals*, (Busnel, R.G. ed), London, Elsevier

Brillard, J.P. and McDaniel, G.R. (1986) Influence of spermatozoa numbers and insemination frequency on fertility in dwarf broiler breeder hens. *Poultry Science*, **65**, 2330–2334

Brillard, J.P., Galut, O. and Nys, Y. (1987) Possible causes of subfertility in hens following insemination near the time of oviposition. *British Poultry Science*, **28**, 307–318

Brocklehurst, D.S. (1975) A preliminary report on a survey of floor laying in breeding stock. Edinburgh, East of Scotland College of Agriculture

Broom, D.M. (1990) Effects of handling and transport on laying hens. *World's Poultry Science Journal*, **46**, 48–50

Brown, E. (1929) *Poultry Breeding and Production*, Vol. 1. London, Ernest Benn

Calet, C. (1965) The relative value of pellets versus mash and grain in poultry nutrition. *World's Poultry Science Journal*, **21**, 23–52

Campos, E.J., Krueger, W.F. and Bradley, J.W. (1971) Maintaining broiler breeders in cages. *Poultry Science*, **50**, 1561

Campos, E.J., Krueger, W.F. and Bradley, J.W. (1973) Performance of commercial broiler breeders in cages. *Poultry Science*, **52**, 2007

Candland, D.K., Taylor, D.B., Dresdale, L., Leiphart, J.M. and Solow, S.P. (1969) Heart rate, aggression and dominance in the domestic chicken. *Journal of Comparative Physiology and Psychology*, **67**, 70–76

Carpenter, G.A., Smith, W.K., MacLaren, A.P.C. and Spackman, D. (1986) Effect of internal air filtration on the performance of broilers and the aerial concentrations of dust and bacteria. *British Poultry Science*, **27**, 471–480

Carter, T.C. (1971) The hen's egg: shell cracking at oviposition and its inheritance. *British Poultry Science*, **12**, 259–278

CEC (Commission of the European Communities) (1985) Amendment 1943/85 to Regulation 95/69, also amended by 927/69 and 2502/71. *Official Journal of the European Communities*, 13th July

CEC (Commission of the European Communities) (1986) Council Directive 86/113/EEC: Welfare of Battery Hens. *Official Journal of the European Communities*, (L 95) **29**, 45–49

CEC (Commission of the European Communities) (1988) Council Directive 88/166/EEC: amendment to 86/113/EEC. *Official Journal of the European Communities*, (L74) 19th March, 83

Cherry, P. and Barwick, M.W. (1962) The effect of light on broiler growth. 1. Light intensity and colour. *British Poultry Science*, **3**, 31

Clayton, G.A., Lake, P.E., Nixey, C., Jones, D.R., Charles, D.R., Hopkins, J.R., Binstead, J.A. and Pickett, R. (1985) *Turkey Production: Breeding and Husbandry*, Ministry of Agriculture, Fisheries and Food reference book 242. London, Her Majesty's Stationery Office

Coenen, A.M.L., Wolters, E.M.T.J., van Luijtelaar, E.L.J.M. and Blokhuis, H.J. (1988) Effects of intermittent lighting on sleep and activity in the domestic hen. *Applied Animal Behaviour Science*, **20**, 309–318

Collias, N.E. and Collias, E.C. (1967) A field study of the Red Jungle Fowl in North-Central India. *Condor*, **69**, 360–386

Collias, N.E., Collias, E.C., Hunsaker, D. and Minning, L. (1966) Locality fixation, mobility and social organization within an unconfined population of Red Jungle Fowl. *Animal Behaviour*, **14**, 550–559

Coltherd, J.B. (1966) The domestic fowl in Ancient Egypt. *Ibis*, **108**, 217–223

Cooper, D.M. (1969) The use of artificial insemination in poultry breeding, the evaluation of semen dilution and storage. In *The Fertility and Hatchability of the Hen's Egg* (Carter, T.C. and Freeman, B.M. eds), pp. 31–44. Edinburgh, Oliver and Boyd

COVP (Centrum voor Onderzoek en Voorlichting voor der Pluimveehouderij) (1988) The tiered wire floor system for laying hens: development and testing of an alternative aviary for laying hens, 1980–1987. *COVP Spelderholt Report* No. 484, Beekbergen, Netherlands

Cowan, P.J., Michie, W. and Roele, D.J. (1978) Choice feeding of the egg type pullet. *British Poultry Science*, **19**, 153–157

Craig, J.V. (1980) Training colony-cage pullets to use nests in mating pens. *Poultry Science*, **59**, 1596

Craig, J.V. (1981) *Domestic Animal Behavior: Causes and Implications for Animal Care and Management*, New Jersey, Prentice Hall

Craig, J.V. and Bhagwat, A.L. (1974) Agonistic and mating behavior of adult chickens modified by social and physical environments. *Applied Animal Ethology*, **1**, 57–65

Craig, J.V., Al-Rawi, B.A. and Kratzer, D.D. (1977) Social status and sex ratio effects on mating frequency of cockerels. *Poultry Science*, **56**, 762–772

Craig, J.V., Biswas, D.K. and Guhl, A.M. (1969) Agonistic behaviour influenced by strangeness, crowding and heredity in female domestic fowl (*Gallus gallus*). *Animal Behaviour*, **17**, 498–506

Curtis, S.E. (1983) *Environmental Management in Animal Agriculture*. Iowa, Iowa State University Press

Cuthbertson, G.J. (1980) Genetic variation in feather pecking behaviour. *British Poultry Science*, **21**, 447–450

Dagnall, S.P. (1989) Use of poultry litter as a fuel. *British Poultry Science*, **30**, 962

Dawkins, M. (1980) *Animal Suffering: the Science of Animal Welfare*. London, Chapman and Hall

Dawkins, M. (1981) Priorities in the cage size and flooring preferences of domestic hens. *British Poultry Science*, **22**, 255–263

Dawkins, M.S. (1983) Battery hens name their price: consumer demand theory and the measurement of ethological needs. *Animal Behaviour*, **31**, 1195–1205

Dawkins, M.S. (1985) Cage height preference and use in battery-kept hens. *Veterinary Record*, **116**, 345–347

Dawkins, M.S. (1988) Behavioural deprivation: a central problem in animal welfare. *Applied Animal Behaviour Science*, **20**, 209–225

Dawkins, M.S. (1989) Time budgets in Red Jungle Fowl as a basis for the assessment of welfare in domestic fowl. *Applied Animal Behaviour Science*, **24**, 77–80

Dawkins, M.S. and Beardsley, T. (1986) Reinforcing properties of access to litter in hens. *Applied Animal Behaviour Science*, **15**, 351–364

Dawkins, M.S. and Hardie, S. (1989) Space needs of laying hens. *British Poultry Science*, **30**, 413–416

Dawkins, M.S. and Nicol, C.J. (1989) No room for manoeuvre. *New Scientist*, 16 September, 44–46

Desforges, M.F. and Wood-Gush, D.G.M. (1975a) A behavioural comparison of mallard and domestic duck. Habituation and flight reactions. *Animal Behaviour*, **23**, 692–697

Desforges, M.F. and Wood-Gush, D.G.M. (1975b) A behavioural comparison of mallard and domestic duck. Spatial relations in small flocks. *Animal Behaviour*, **23**, 698–705

Dorminey, R.M. (1974) Incidence of floor eggs as influenced by time of nest installation, artificial lighting and nest location. *Poultry Science*, **53**, 1886–1891

Duff, S.R.I., Hocking, P.M., Randall, C.J. and MacKenzie, G. (1989) Head swelling of traumatic aetiology in broiler breeding fowl. *Veterinary Record*, **125**, 133–134

Duncan, E.T., Appleby, M.C. and Hughes, B.O. (1992) Effect of perches in laying cages on welfare and production of laying. *British Poultry Science* , **33**, 25–35

Duncan, I.J.H. (1970) Frustration in the fowl. In *Aspects of Poultry Behaviour* (Freeman B.M. and Gordon, R.F. eds), pp. 15–31. Edinburgh, British Poultry Science

Duncan, I.J.H. (1978) An overall assessment of poultry welfare. In *Proceedings, First Danish Seminar on Poultry Welfare* (Sorensen, L.Y. ed.). Copenhagen, World's Poultry Science Association

Duncan, I.J.H. (1982) Investigations into the feelings of the domestic fowl: what's all the flap about? *CEC Research on Poultry Welfare*. Progress Reports 1981/1982

Duncan, I.J.H. and Hughes, B.O. (1988) Can the welfare needs of poultry be measured? In *Science and the Poultry Industry* (Hardcastle, J. ed.), pp. 24–25. London, Agricultural and Food Research Council

Duncan, I.J.H. and Kite, V. (1987) Some investigations into motivation in the domestic fowl. *Applied Animal Behaviour Science*, **18**, 387–388

Duncan, I.J.H. and Kite, V.G. (1989) Nest site selection and nest building behaviour in domestic fowl. *Animal Behaviour*, **37**, 215–231

Duncan, I.J.H. and Petherick, J.C. (1989) Cognition: the implication for animal welfare. *Applied Animal Behaviour Science*, **24**, 81

Duncan, I.J.H., Hocking, P.M. and Seawright, E. (1990) Sexual behaviour and fertility in broiler breeder domestic fowl. *Applied Animal Behaviour Science*, **26**, 201–213

Duncan, I.J.H., Savory, C.J. and Wood-Gush, D.G.M. (1978) Observations on the reproductive behaviour of domestic fowl in the wild. *Applied Animal Ethology*, **4**, 29–42

Ede, D.A. (1964) *Bird Structure. An Approach through Evolution, Development and Function in the Fowl*. London, Hutchinson Educational Ltd

Ehlhardt, D.A. and Koolstra, C.L.M. (1984) Multi-tier system for housing laying hens. *Pluimveehouderij*, 21st December, 44–47

Elson, H.A. (1979) Design of equipment for feeding the bird. In *Poultry Science*

Symposium, No. 14 (Boorman, K.N. and Freeman, B.M. eds), pp. 431–444. Edinburgh, British Poultry Science Limited

Elson, H.A. (1981) Modified cages for layers. In *Alternatives to Intensive Husbandry Systems*, pp. 47–50, Potters Bar, Universities Federation for Animal Welfare

Elson, H.A. (1985) The economics of poultry welfare. In *Proceedings, Second European Symposium on Poultry Welfare* (Wegner, R.M. ed.), pp. 244–253, Celle, World's Poultry Science Association

Elson, H.A. (1988a) Making the best cage decisions. In *Cages for the Future*, pp. 70–76. Cambridge Poultry Conference, Agricultural Development and Advisory Service

Elson, H.A. (1988b) Walk-about cages on test. *Poultry World*, **142**, (5), 1–4

Elson, H.A. (1989) Improvements in alternative systems of egg production. In *Proceedings, Third European Symposium on Poultry Welfare* (Faure, J.M. and Mills, A.D. eds), pp. 183–199. Tours, World's Poultry Science Association

Elson, H.A. (1990) Recent developments in laying cages designed to improve bird welfare. *World's Poultry Science Journal*, **46**, 34–37

Emmans, G.C. (1975) Problems in feeding laying hens: can a system based on choice solve them? *World's Poultry Science Journal*, **31**, 31

Emmans, G.C. (1977) The nutrient intake of laying hens given a choice of diets in relation to their production requirements. *British Poultry Science*, **18**, 227–236

Emmans, G.C. and Charles, D.R. (1977) Climatic environment and poultry feeding in practice. In *Nutrition and the Climatic Environment* (Haresign, W., Swan, H. and Lewis, D., eds), pp. 31–49. London, Butterworths

Engelmann, C. (1940) Versuche uber die 'Beliebtheit' einiger Getreidearten beim Huhn. *Zeitschrift fur vergleichende Physiologie*, **27**, 526–634

Esmay, M.L. (1978) *Principles of Animal Environment*. Westport, AVI

Everton, A. (1989) The legal protection of farm livestock: avoidance of 'unnecessary suffering' and the positive promotion of welfare. In *Animal Welfare and the Law*, (Blackman, D.E., Humphries P.N. and Todd, P. eds). Cambridge, Cambridge University Press

Farm Animal Welfare Council (1986) *An Assessment of Egg Production Systems*. Tolworth, Farm Animal Welfare Council

Faure, J.M. (1980) To adapt the environment to the bird or the bird to the environment? In *The Laying Hen and its Environment* (Moss, R. ed.), pp. 19–30. The Hague, Martinus Nijhoff

Faure, J.M. and Jones, R.B. (1982) Effects of age, access and time of day on perching behaviour in the domestic fowl. *Applied Animal Ethology*, **8**, 357–364

Fischer, G.J. (1975) The behaviour of chickens. In *The Behaviour of Domestic Animals*, 3rd edn. (Hafez, E.S.E. ed.), pp. 454–489. London, Bailliere Tindall

Freeman, B.M. (1983) Floor space allowance for the caged domestic fowl. *Veterinary Record*, **112**, 562–563

Freeman, B.M. (1983–84) *Physiology and Biochemistry of the Domestic Fowl*, Vols 4, 5. London, Academic Press

Fujita, H. (1973) Quantitative studies on the variations in feeding activity of chickens. II. Effect of the physical form of the feed on the feeding activity of laying hens. *Japanese Poultry Science*, **10**, 47–54

Gentle, M.J. (1975) Gustatory behaviour of the chicken and other birds. In *Neural and Endocrine Aspects of Behaviour in Birds* (Wright, P., Caryl, P.G. and Vowles, D.M. eds). Amsterdam Elsevier

Gentle, M.J. (1986a) Beak trimming in poultry. *World's Poultry Science Journal*, **42**, 268–275

Gentle, M.J. (1986b) Aetiology of food-related oral lesions in chickens. *Research in Veterinary Science*, **40**, 219–224

Gentle, M.J., Hughes, B.O. and Hubrecht, R.C. (1982) The effect of beak trimming on food intake, feeding behaviour and body weight in adult hens. *Applied Animal Ethology*, **8**, 147–159

Gerken, M. and Petersen, J. (1987) Bidirectional selection for dustbathing activity in Japanese quail (*Coturnix coturnix japonica*). *British Poultry Science*, **28**, 23–37

Gibson, S.W., Dun, P. and Hughes, B.O. (1988) The performance and behaviour of laying fowls in a covered strawyard system. *Research and Development in Agriculture*, **5**, 153–163

Gibson, S.W., Innes, J. and Hughes, B.O. (1985) Aggregation behaviour of laying fowls in a covered strawyard. In *Proceedings, Second European Symposium on Poultry Welfare* (Wegner, R.M. ed.), pp. 295–298. Celle, World's Poultry Science Association

Goodale, H.D., Sanborn, R. and White, D. (1920) Broodiness in domestic fowl. *Bulletin of Massachusetts Agricultural Experimental Station*, 199

Gray, P.H. and Howard, K.I. (1957) Specific recognition of humans in imprinted chicks. *Perceptual and Motor Skills*, **7**, 301–304

Gregory, N.G. and Wilkins, L.J. (1989) Broken bones in domestic fowl: handling and processing damage in end-of-lay battery hens. *British Poultry Science*, **30,** 555–562

Gregory, N.G., Wilkins, L.J., Eleperuma, S.D., Ballantyne, A.J. and Overfield, N.D. (1990) Broken bones in domestic fowls: effects of husbandry system and stunning method in end-of-lay hens. *British Poultry Science*, **31**, 59–69

Guhl, A.M. (1953) Social behaviour of the domestic fowl. *Technical Bulletin of the Kansas Agricultural Experiment Station*, **73**, 48pp.

Guhl, A.M., Collias, N.E. and Allee, W.C. (1945) Mating behavior and the social hierarchy in small flocks of White Leghorns. *Physiological Zoology*, **18**, 365–390

Guyomarc'h, C., Guyomarc'h, J.C. and Garnier, D.H. (1981) Influence of male vocalisations on the reproduction of quail females (*Coturnix coturnix japonica*). *Biology of Behaviour*, **6**, 167–182

Haartsen, P.I. and Elson, H.A. (1989) Economics of alternative housing systems. In *Alternative Improved Housing Systems for Poultry*, pp. 143–150. CEC Seminar, EUR. 11711, Beekbergen

Hale, E.B. (1955) Defects in sexual behavior as factors affecting fertility in turkeys. *Poultry Science*, **34**, 1059–1067

Hale, E.B. (1975) Domestication and the evolution of behaviour. In *The Behaviour of Domestic Animals*, 3rd edn (Hafez, E.S.E. ed.). London, Baillière Tindall

Hale, E.B., Schleidt, W.M. and Schein, M.W. (1975) The behaviour of turkeys. In *The Behaviour of Domestic Animals*, 3rd edn (Hafez, E.S.E. ed.). London, Baillière Tindall

Halpern, B.P. (1962) Gustatory nerve responses in the chicken. *American Journal of Physiology*, **203**, 541–544

Hann, C.M. (1980) Some system definitions and characteristics. In *The Laying Hen and Its Environment* (Moss, R. ed.), pp. 239–258. The Hague, Martinus Nijhoff

Hansen, R.S. (1976) Nervousness and hysteria of mature female chickens. *Poultry Science*, **55**, 531–543

Harrison, R. (1964) *Animal Machines*. London, Stuart

Harrison, R. (1989) Research into action – some concerns. In *Proceedings, Third European Symposium on Poultry Welfare* (Faure, J.M. and Mills, A.D. eds), pp. 253–255. Tours, World's Poultry Science Association

Harvey, S. and Bedrak, E. (1984) Endocrine basis of broodiness in poultry. In *Reproductive Biology of Poultry*, (Cunningham, F.J., Lake, P.E. and Hewitt, D. eds), pp. 111–132. Harlow, British Poultry Science Ltd

Health and Safety Executive (1980) *Threshold Limit Values*. London, Her Majesty's Stationery Office

Hearn, P.J. (1976) A comparison of troughs, nipples and cup drinkers for laying hens in cages. In *Gleadthorpe Experimental Husbandry Farm Poultry Booklet*, pp. 94–98

Hearn, P.J. (1981) The effect of time of feeding and position of nest boxes on floor eggs. MAFF/ADAS Report PH 03555

Hearn, P.J. and Gooderham, K.R. (1988) Ducks. In *Management and Welfare of Farm Animals* (Universities Federation for Animal Welfare, eds), pp. 243–253. London, Baillière Tindall

Heil, G., Simianer, H. and Dempfle, L. (1990) Genetic and phenotypic variation in prelaying behaviour of Leghorn hens kept in single cages. *Poultry Science*, **69**, 1231–1235

Hewson, P. (1986) Origin and development of the British poultry industry: the first hundred years. *British Poultry Science*, **27**, 525–539

Hill, J.A. (1977) 'The Relationship Between Food and Water Intake in the Laying Hen'. PhD Thesis, Huddersfield Polytechnic

Hill, J.A. (1983) Aviary system poses feather pecking and floor egg problems. *Poultry International*, May, 109–113

Hill, J.A. (1986) Egg production in alternative systems – a review of recent research in the UK. *Research and Development in Agriculture*, **3**, 13–18

Hill, J.A., Powell, A.J. and Charles, D.R. (1979) Water intake. In *Food Intake Regulation in Poultry* (Boorman K.N. and Freeman, B.M. eds), pp. 231–257. Edinburgh, British Poultry Science Ltd

HMSO (Her Majesty's Stationery Office) (1965) *Report of the Technical Committee to Enquire into the Welfare of Animals kept under Intensive Livestock Husbandry Systems*. Command Paper 2836. London, Her Majesty's Stationery Office

HMSO (Her Majesty's Stationery Office) (1987) *Animals, Prevention of Cruelty, The Welfare of Battery Hens Regulations 1987*. London, Her Majesty's Stationery Office

Hocking, P.M., Waddington, D., Walker, M.A. and Gilbert, A.B. (1989) Control of the development of the ovarian follicular hierarchy in broiler breeder pullets by food restriction during rearing. *British Poultry Science*, **30**, 161–174

Hodgetts, B. (1981) *Dealing with Dirty Hatching Eggs*. MAFF Information for Flock Farms and Hatcheries: Hatch Handout 17

Hogan, J.A. (1971) The development of a hunger system in young chicks. *Behaviour*, **39**, 128–201

Hogan, J.A. (1973) Development of food recognition in young chicks. 1. Maturation and nutrition. *Journal of Comparative and Physiological Psychology*, **83**, 355–366

Holcombe, D.J., Roland, D.A. and Harms, R.H. (1976a) The ability of hens to regulate phosphorus intake when offered diets containing different levels of phosphorus. *Poultry Science*, **55**, 308–317

Holcombe, D.J., Roland, D.A. and Harms, R.H. (1976b) The ability of hens to regulate protein intake when offered a choice of diets containing different levels of protein. *Poultry Science,* **55**, 1731–1737

Huber, H., Fölsch, D.W. and Stahli, U. (1985) Influence of various nesting materials on nest-site selection of the domestic hen. *British Poultry Science,* **26**, 367–373

Hughes, B.O. (1971) Allelomimetic feeding in the domestic fowl. *British Poultry Science,* **12**, 359–366

Hughes, B.O. (1972) A circadian rhythm of calcium intake in the domestic fowl. *British Poultry Science,* **13**, 485–493

Hughes, B.O. (1973) The effect of implanted gonadal hormones on feather pecking and cannibalism in pullets. *British Poultry Science,* **14**, 341–348

Hughes, B.O. (1975a) Spatial preference in the domestic hen. *British Veterinary Journal,* **131**, 560–564

Hughes, B.O. (1975b) The concept of an optimal stocking density and its selection for egg production. In *Economic Factors Affecting Egg Production* (Freeman, B.M. and Boorman, K.N. eds), *Poultry Science Symposium,* **10**, 271–298. Edinburgh, British Poultry Science Ltd

Hughes, B.O. (1976) Preference decisions of domestic hens for wire or litter floors. *Applied Animal Ethology,* **2**, 155–165

Hughes, B.O. (1977) Selection of group size by individual laying hens. *British Poultry Science,* **18**, 9–18

Hughes, B.O. (1979) Appetites for specific nutrients. In *Food Intake Regulation in Poultry* (Boorman, K.N. and Freeman, B.M. eds), pp. 141–169. Edinburgh, British Poultry Science Ltd

Hughes, B.O. (1983a) Conventional and shallow cages: a summary of research from welfare and production aspects. *World's Poultry Science Journal,* **39**, 218–228

Hughes, B.O. (1983b) Headshaking in fowls: the effect of environmental stimuli. *Applied Animal Ethology,* **11**, 45–53

Hughes, B.O. (1984) The principles underlying choice feeding behaviour in fowls – with special reference to production experiments. *World's Poultry Science Journal,* **40**, 141–150

Hughes, B.O. (1985) Feather loss as a problem: how does it occur? In *Proceedings, Second European Seminar on Poultry Welfare* (Wegner, R.M. ed.), pp. 177–188. Celle, World's Poultry Science Association.

Hughes, B.O. and Appleby, M.C. (1989) Increase in bone strength of spent laying hens housed in modified cages with perches. *Veterinary Record,* **124**, 483–484

Hughes, B.O. and Black, A.J. (1973) The preference of domestic hens for different types of battery cage floor. *British Poultry Science,* **14**, 615–619

Hughes, B.O. and Black, A.J. (1974) The effect of environmental factors on activity, selected behaviour patterns and 'fear' of fowls in cages and pens. *British Poultry Science,* **15**, 375–380

Hughes, B.O. and Black, A.J. (1976) Battery cage shape: its effect on diurnal feeding pattern, egg shell cracking and feather pecking. *British Poultry Science,* **17**, 327–336

Hughes, B.O. and Dewar, W.A. (1971) A specific appetite for zinc in zinc-depleted domestic fowl. *British Poultry Science,* **12**, 255

Hughes, B.O. and Dun, P. (1983) A comparison of laying stock housed intensively in cages and outside on range. Years 1981–1983. *Research and Development Publication No. 18.* Ayr, West of Scotland Agricultural College

Hughes, B.O. and Dun, P. (1986) A comparison of hens housed intensively in cages and outside on range. *Zootechnica International*, February, 44–46

Hughes, B.O. and Duncan, I.J.H. (1972) The influence of strain and environmental factors upon feather pecking and cannibalism in fowls. *British Poultry Science*, **13**, 525–547

Hughes, B.O. and Duncan, I.J.H. (1988) The notion of ethological 'need', models of motivation and animal welfare. *Animal Behaviour*, **36**, 1696–1707

Hughes, B.O. and Michie, W. (1982) Plumage loss in medium bodied hybrid hens: the effect of beak trimming and cage design. *British Poultry Science*, **23**, 59–64

Hughes, B.O. and Whitehead, C.C. (1979) Behavioural changes associated with the feeding of low-sodium diets to laying hens. *Applied Animal Ethology*, **5**, 255–266

Hughes, B.O. and Wood-Gush, D.G.M. (1971) Investigations into specific appetites for sodium and thiamine in domestic fowls. *Physiology and Behavior*, **6**, 331–339

Hughes, B.O. and Wood-Gush, D.G.M. (1977) Agonistic behaviour in domestic hens: the influence of housing method and group size. *Animal Behaviour*, **25**, 1056–1062

Hughes, B.O., Duncan, I.J.H. and Brown, M.F. (1989) The performance of nest building by domestic hens: is it more important than the construction of a nest? *Animal Behaviour*, **37**, 210–214

Hughes, B.O., Gilbert, A.B. and Brown, M.F. (1986) Categorisation and causes of abnormal egg shells: relationship with stress. *British Poultry Science*, **27**, 325–337

Hughes, B.O., Wilson, S.C., Appleby, M.C. and Smith, S.F. (1992) Effect of perches on bone structure and strength in caged laying hens. *Research in Veterinary Science* (in press)

Huon, F., Meunier-Salaun, M-C. and Faure, J-M. (1986) Feeder design and available feeding space influence the feeding behaviour of hens. *Applied Animal Behaviour Science,* **15**, 65–70

Jackson, C. (1989) Europe and animal welfare. In *Animal Welfare and the Law* (Blackman, D.E. Humphries, P.N. and Todd. P. eds). Cambridge, Cambridge University Press

Jackson, W.T. (1980) 'Laws and Other Measures for the "On-farm" Protection of Farm Animals in the Member States of the Council of Europe'. Thesis, University of Berne

Jackson, W.T. (1989) On-farm animal welfare law in Europe – using the law. *Applied Animal Behaviour Science*, **20**, 165–173

Jenner, T.D. and Appleby, M.C. (1991) Effect of space allowance on behavioural restriction and synchrony in hens. *Applied Animal Behaviour Science*, **31**, 292–293

Jensen, J.F. (1981) Induced moulting. In *Proceedings, First European Symposium on Poultry Welfare* (Sorensen, L.Y. ed.), pp. 165–173. Copenhagen, World's Poultry Science Association

Jones, J.E., Wilson, H.R., Harms, R.H., Simpson, C.F. and Waldroup, P.W. (1967) Reproductive performance in male chickens fed protein deficient diets during the growing period. *Poultry Science*, **46**, 1569–1577

Jones, M.C. and Leighton, A.T., Jr. (1987) Research note: effect of presence or absence of the opposite sex on egg production and semen quality of breeder turkeys. *Poultry Science*, **66**, 2056–2059

Jones, R.B. (1982) Effects of early environmental enrichment upon open field behavior and timidity in the domestic chick. *Developmental Psychobiology*, **15**, 105–111

Jones, R.B. (1987a) Fearfulness of caged laying hens: the effects of cage level and type of roofing. *Applied Animal Behaviour Science*, **17**, 171–175

Jones, R.B. (1987b) Social and environmental aspects of fear in the domestic fowl. In *Cognitive Aspects of Social Behaviour in the Domestic Fowl* (Zayan, R. and Duncan, I.J.H. eds). Amsterdam, Elsevier

Jones, R.B. (1989) Development and alleviation of fear. In *Proceedings of the Third European Symposium on Poultry Welfare* (Faure, J.M. and Mills, A.D. eds), pp. 123–136. Tours, World's Poultry Science Association

Jones, R.B., and Gentle, M.J. (1985) Olfaction and behavioural modification in domestic chicks (*Gallus domesticus*). *Physiology and Behavior*, **34**, 917–924

Jones, R.B., Duncan, I.J.H. and Hughes, B.O. (1981) The assessment of fear in domestic hens exposed to a looming human stimulus. *Behavioural Processes*, **6**, 121–133

Jones, R.B., Mills, A.D. and Faure, J.M. (1991) Genetic and experiential manipulation of fear-related behavior in Japanese quail (*Coturnix c. japonica*). *Journal of Comparative Psychology*, **105**, 15–24

Justice, W.P., McDaniel, G.R. and Craig, J.V. (1962) Techniques for measuring sexual effectiveness in male chickens. *Poultry Science*, **41**, 732–739

Keeling, L.J. and Duncan, I.J.H. (1989) Interindividual distances and orientation in laying hens housed in groups of three in two different sized enclosures. *Applied Animal Behaviour Science*, **24**, 325–342

Keeling, L.J., Hughes, B.O. and Dun, P. (1988) Performance of free range laying hens in a polythene house and their behaviour on range. *Farm Buildings Progress*, **94**, 21–28

King, A.S. and McLelland, J. (1975) *Outline of Avian Anatomy*, London, Baillière Tindall

King, A.S. and McLelland, J. (1979–85) *Form and Function in Birds*, Vols 1–3. London, Academic Press

King, A.W.M. and Dun, P. (1984) Personal communication regarding ventilated litter floor colony system for layers, cited by Elson, 1985

Kirk, S., Emmans, G.C., McDonald, R. and Arnist, D. (1980) Factors affecting the hatchability of eggs from broiler breeders. *British Poultry Science*, **21**, 37–43

Kite, V.C., Cumming, R.B. and Wodzicka-Tomaszewska, M. (1980) Nesting behaviour of hens in relation to the problem of floor eggs. In *Behaviour in Relation to Reproduction, Management and Welfare of Farm Animals* (Wodzicka-Tomaszewska, M., Edey, T.N. and Lynch, J.J. eds), pp. 93–96. Reviews in Rural Science IV, Armidale, Australia

Klopfer, P.H. (1959) An analysis of learning in young Anatidae. *Ecology*, **40**, 90–102

Knowles, T.G. and Broom, D.M. (1990) Limb bone strength and movement in laying hens from different housing systems. *Veterinary Record*, **126**, 354–356

Kratzer, D.D. and Craig, J.V. (1980) Mating behavior of cockerels: effects of social status, group size and group density. *Applied Animal Ethology*, **6**, 49–62

Kruijt, J.P. (1964) *Ontogeny of Social Behaviour in Burmese Red Junglefowl*, Leiden Brill

Lake, P.E. (1969) Factors affecting fertility. In *The Fertility and Hatchability of the Hen's Egg* (Carter, T.C. and Freeman, B.M. eds), pp. 3–29. Edinburgh, Oliver and Boyd

Landauer, W. (1967) The hatchability of chicken eggs as influenced by environment and heredity. *Storrs Agricultural Experimental Station Monographs 1* (revised)

Lea, R.W., Dods, A.S.M., Sharp, P.J. and Chadwick, A. (1981) The possible role of prolactin in the regulation of nesting behaviour and the secretion of luteinizing hormone in broody bantams. *Journal of Endocrinology*, **91**, 89–99

Lewis, P.D., Perry, G.C. and Tuddenham, A. (1987) Noise output of hens subjected to interrupted lighting regimens. *British Poultry Science*, **28**, 535–540

Lintern-Moore, S. (1972) The relationship between water intake and the production of 'wet' droppings in the domestic fowl. *British Poultry Science*, **13**, 237–242

Lorenz, F.W. (1959) Reproduction in the domestic fowl: physiology of the male. In *Reproduction in Domestic Animals*, Vol. 2 (Cole, H.H. and Cupps, P.T. eds), pp. 343–398. New York, Academic Press

Lynn, N.J. (1989) Effect of the degree and duration of different feeding regimens on laying hens undergoing an induced moult. *British Poultry Science*, **30**, 970

MAFF (Ministry of Agriculture, Fisheries and Food) (1970) *Welfare of Livestock. Codes of Recommendations for the Welfare of Livestock (Cattle, Pigs, Domestic Fowls and Turkeys) A Report by the Farm Animal Welfare Advisory Committee.* Tolworth, Animal Health Division

MAFF (Ministry of Agriculture, Fisheries and Food) (1987a) *Codes of Recommendations for the Welfare of Livestock: Domestic Fowls.* London, MAFF Publications Office

MAFF (Ministry of Agriculture, Fisheries and Food) (1987b) *Codes of Recommendations for the Welfare of Livestock: Ducks.* London, MAFF Publications Office

MAFF (Ministry of Agriculture, Fisheries and Food) (1987c) Statutory Instrument 1987 No. 2020. *The Welfare of Battery Hens Regulations 1987.* London, Her Majesty's Stationery Office

MAFF (Ministry of Agriculture, Fisheries and Food) (1988) Agricultural Statistics for the United Kingdom. London, Her Majesty's Stationery Office

Maguire, S.N. (1986) 'A Study of Some Factors Influencing the Nest-site Selection of the Domestic Hen in Relation to the Problem of Floor Eggs', MSc thesis, University of New England, Australia

Manning, A. (1967) *An Introduction to Animal Behaviour*, London, Edward Arnold

Marsden A., Morris, T.R. and Cromarty, A.S. (1987) Effects of constant environmental temperatures on the performance of laying pullets. *British Poultry Science*, **28**, 361–380

Mashaly, M.M., Webb, M.L., Youtz, S.L., Roush, W.B. and Graves, H.B. (1984) Changes in serum corticosterone concentration of laying hens as a response to increased population density. *Poultry Science*, **63**, 2271–2274

Mastika, M. and Cumming, R.B. (1981) Performance of two strains of broiler chickens offered free choice from different ages. *Proceedings of the 4th Australasian Poultry Stock Feed Convention*, p. 74

Matter, F. and Oester, H. (1989) Hygiene and welfare implications of alternative husbandry systems for laying hens. In *Proceedings, Third European Symposium on Poultry Welfare* (Faure, J.M. and Mills, A.D. eds), pp. 201–212, Tours, World's Poultry Science Association

Maxwell, M.H., Dick, L.A., Anderson, I.A. and Mitchell, M.A. (1989) Ectopic cartilaginous and osseous lung nodules induced in the young broiler by inadequate ventilation. *Avian Pathology*, **18**, 113–124

May, G.C. and Hawksworth, D. (1982) *British Poultry Standards*, 4th edn. London, Butterworth Scientific

McBride, G. (1958) Relationship between aggressiveness and egg production in the domestic hen. *Nature*, **181**, 858

McBride, G. (1970) The social control of behaviour in fowls. In *Aspects of Poultry*

Behaviour (Freeman, B.M. and Gordon, R.F. eds), pp. 3–13. Edinburgh, British Poultry Science

McBride, G. and Foenander, F. (1962) Territorial behaviour in flocks of domestic fowls. *Nature, London*, **194,** 102

McBride, G., Parer, I.P. and Foenander, F. (1969) The social organization and behaviour of the feral domestic fowl. *Animal Behaviour Monographs*, **2**, 125–181

McDaniel, G.R. and Craig, J.V. (1959) Behavior traits, semen measurements and fertility of White Leghorn males. *Poultry Science*, **38**, 1005–1014

McLean, K.A., Baxter, M.R. and Michie, W. (1986) A comparison of the welfare of laying hens in battery cages and in a perchery. *Research and Development in Agriculture*, **3**, 93–98

Meijsser, F.M. and Hughes, B.O. (1989) Comparative analysis of pre-laying behaviour in battery cages and in three alternative systems. *British Poultry Science*, **30**, 747–760

Michie, W. and Wilson, C.W. (1984) The perchery system of housing commercial layers. *World's Poultry Science Journal*, **40**, 179

Michie, W. and Wilson, C.W. (1985) *The Perchery System for Housing Layers*. Scottish Agricultural Colleges Research and Development Note No. 25

Mills, A.D., Wood-Gush, D.G.M. and Hughes, B.O. (1985a) Genetic analysis of strain differences in pre-laying behaviour in battery cages. *British Poultry Science*, **26**, 182–197

Mills, A.D., Duncan, I.J.H., Slee, G.S. and Clark, J.S.B. (1985b) Heart rate and laying behavior in two strains of domestic chicken. *Physiology of Behavior*, **35**, 145–147

Mongin, P. and Saveur, B. (1974) Voluntary food and calcium intake by the laying hen. *British Poultry Science*, **15**, 349–359

Mongin, P. and Saveur, B. (1979) The specific calcium appetite of the domestic fowl. In *Food Intake Regulation in Poultry* (Boorman K.N. and Freeman, B.M. eds), pp. 171–189. Edinburgh, British Poultry Science Ltd

MORI (Market and Opinion Research International) (1983) *Public Attitudes towards Farmers. Research Study Conducted for National Farmers' Union*. London, Market and Opinion Research International

Morris, T. R. (1967) Light requirements of the fowl. In *Environmental Control in Poultry Production* (Carter, T.C., ed.), pp. 15–39. Edinburgh, Oliver and Boyd

Morris, T.R. and Bhatti, T.M. (1978) Entrainment of oviposition in the fowl using bright and dim light cycles. *British Poultry Science*, **19**, 341–348

Morris, T.R., Midgley, M. and Butler, E.A. (1988) Experiments with Cornell intermittent lighting systems for laying hens. *British Poultry Science*, **29**, 325–332

Neuhaus, W. (1963) On the olfactory sense of birds. In *Olfaction and Taste* (Zotterman, Y. ed.). Oxford, Pergamon

NFU (National Farmers' Union) (1990) *Poultry, Some Facts and Figures*. London, National Farmers Union

Nicol, C.J. (1987a) Effect of cage height and area on the behaviour of hens housed in battery cages. *British Poultry Science*, **28**, 327–335

Nicol, C.J. (1987b) Behavioural responses of laying hens following a period of spatial restriction. *Animal Behaviour*, **35**, 1709–1719

Nicol, C.J. and Dawkins, M.S. (1990) Homes fit for hens. *New Scientist*, 17th March, 46–51

NOP (National Opinion Polls) (1983) *Animal Issues and their Influence on Voting*. London, National Opinion Polls Market Research Ltd

Nørgaard-Nielsen, G. (1986) Behaviour, health and production of egg-laying hens in the Hans Kier system compared to hens on litter and in battery cages. *Rapport til Hans Kier Fond, Forseningen til Dyrenes Beskyttelsei Danmark*, 1–198

Nørgaard-Nielsen, G. (1990) Bone strength of laying hens kept in an alternative system, compared with hens in cages and on deep-litter. *British Poultry Science*, **31**, 81–89

Odum, T.W., Wideman, R.F. and Coello, C.L. (1987) Current research on body fluid accumulation in broilers (ascites) *Zootechnica International*, August, 53–54

Oester, H. (1986) Systemes de détention récents pour pondeuses en Suisse. In *Proceedings, 7th European Poultry Conference* (Larbier, M. ed.), pp. 1077–1081. Paris, World's Poultry Science Association

Oester, H. and Frohlich, E. (1989) Alternative systems in Switzerland. In *Alternative Improved Housing Systems for Poultry*, pp. 50–58. CEC Seminar, Beekbergen

Ottinger, M.A. and Brinkley, H.J. (1979) The ontogeny of crowing and copulatory behaviour in Japanese quail (*Coturnix coturnix japonica*). *Behavioural Processes*, **4**, 43–51

Ottinger, M.A. and Mench, J.A. (1989) Reproductive behaviour in poultry: implications for artificial insemination technology. *British Poultry Science*, **30**, 431–442

Ottinger, M.A. Schleidt, W.M. and Rusek, E. (1982) Daily patterns in courtship and mating behavior in the male Japanese quail. *Behavioural Processes*, **7**, 223–233

Oyetunde, O.O.F., Thomson R.G. and Carlson, H.C. (1978) Aerosol exposure of ammonia, dust and *Escherichia coli* in broiler chickens. *Canadian Veterinary Journal*, **19**, 167–193

Pamment, P., Foenander, F. and McBride, G. (1983) Social and spatial organization of male behaviour in mated domestic fowl. *Applied Animal Ethology*, **9**, 341–349

Parker, J.E. and Bernier, P.E. (1950) Relation of male to female ratio in New Hampshire breeder flocks to fertility of eggs. *Poultry Science*, **29**, 377–380

Parker, J.E. Mckenzie, F.F. and Kempster, H.L. (1940) Observations on the sexual behavior of New Hampshire males. *Poultry Science*, **19**, 191–197

Perry, G.C., Charles, D.R., Day, P.J., Hartland, J.R. and Spencer, P.G. (1971) Egg laying behaviour in a broiler parent flock. *World's Poultry Science Journal*, **27**, 162

Perry, G.C., Stevens, K. and Allen, J. (1976) Particle selection by caged layers and pullets. *Proceedings Vth European Poultry Conference, Malta*, pp. 1089–1096. World Poultry Science Association

Politiek, R.D. and Bakker, J.J. eds (1982) *Livestock Production*. Amsterdam, Elsevier

Preston, A.P. and Mulder, J. (1989) Effect of vertical food-trough dividers on the feeding and agonistic behaviour of layer hens. *British Poultry Science*, **30**, 489–496

Preston, A.P. and Murphy, L.B. (1989) Movement of broiler chickens reared in commercial conditions. *British Poultry Science*, **30**, 519–532

Rappaport, S. and Soller, M. (1966) Mating behaviour, fertility and rate-of-gain in Cornish males. *Poultry Science*, **45**, 997–1003

Raud, H. (1990) 'Contribution a l'Etude du Comportement Sexuel du Canard de Barbarie (*Cairina moschata*), PhD thesis, University of Rennes

Rheingold, H.L. and Hess, E.H. (1957) The chick's preference for some visual properties of water. *Journal of Comparative and Physiological Psychology*, **50**, 417–421

Rietveld-Piepers, B., Blokhuis, H.J. and Wiepkema, P.R. (1985) Egg-laying behaviour and nest-site selection of domestic hens kept in small floor pens. *Applied Animal Behaviour Science*, **14**, 75–88

Robertson, E.S., Appleby, M.C., Hogarth, G.S. and Hughes, B.O. (1989) Modified cages for laying hens: a pilot trial. *Research and Development in Agriculture*, **6**, 107–114

Rogers, C.S., Appleby, M.C., Keeling, L., Robertson, E.S. and Hughes, B.O. (1989) Assessing public opinion on commercial methods of egg production: a pilot study. *Research and Development in Agriculture*, **6**, 19–24

Roush, W.B., Mashaly, M.M. and Graves, H.B. (1984) Effect of increased bird population in a fixed cage area on production and economic responses of Single Comb White Leghorn laying hens. *Poultry Science*, **63**, 45–48

Rowland, K.W. (1985) Intermittent lighting for laying fowls: a review. *World's Poultry Science Journal*, **41**, 5–19

Rowland, L.O., Fry, J.L., Christmas, R.B., O'Sheen, A.W. and Harris, R.H. (1972) Differences in tibia strength and bone ash among strains of layers. *Poultry Science*, **51**, 1612–1615

Rozin, P. (1976) The selection of foods by rats, humans and animals. In *Advances in the Study of Behaviour* (Rosenblatt, J.S., Hinde, R.A., Shaw, E. and Beer, C. eds), **6**, 21. New York, Academic Press

Rutter, S.M. and Duncan, I.J.H. (1989) Behavioural measures of aversion in domestic fowl. In *Proceedings, Third European Symposium on Poultry Welfare* (Faure, J.M. and Mills, A.D. eds), pp. 277–279. Tours, World's Poultry Science Association

Rutter, S.M. and Duncan, I.J.H. (1992) Shuttle and one-way avoidance as measures of aversion in the domestic fowl. *Applied Animal Behaviour Science* (in press)

Sainsbury, D.W.B. (1971) Domestic fowls. In *The UFAW Handbook on the Care and Management of Farm Animals*, pp. 99–130. Edinburgh and London. Churchill and Livingstone

Sainsbury, D.W.B. (1980) *Poultry Health and Management*. London, Granada

Savory, C.J. (1974) Growth and behaviour of chicks fed on pellets or mash. *British Poultry Science*, **15**, 281–286

Savory, C.J. (1976) Effects of different lighting regimes on diurnal feeding patterns of the domestic fowl. *British Poultry Science*, **17**, 341–350

Savory, C. J. (1979) Feeding behaviour. In *Food Intake Regulation in Poultry* (Boorman, K.N. and Freeman B.M., eds), pp. 277–323. Edinburgh, British Poultry Science Ltd

Savory, C.J. (1980) Meal occurrence in Japanese quail in relation to particle size and nutrient density. *Animal Behaviour*, **28**, 160–171

Savory, C.J. (1982) Effects of broiler companions on early performance of turkeys. *British Poultry Science*, **23**, 81–88

Savory, C.J. (1989a) Stereotyped behaviour as a coping strategy in restricted-fed broiler breeder stock. In *Proceedings, Third European Symposium on Poultry Welfare* (Faure, J.M. and Mills, A.D. eds), pp. 261–264. Tours, World's Poultry Science Association

Savory, C.J. (1989b) The importance of invertebrate food to chicks of gallinaceous species. *Proceedings of the Nutrition Society*, **48**, 113–133

Savory, C.J., Gentle, M.J. and Yeomans, M.R. (1989) Opioid modulation of feeding and drinking in fowls. *British Poultry Science*, **30**, 379–392

Savory, C.J., Seawright, E. and Watson, A. (1992) Stereotyped behaviour in broiler breeders in relation to husbandry and opioid receptor blockade. *Applied Animal Behaviour Science* (in press)

Savory, C.J., Wood-Gush, D.G.M. and Duncan, I.J.H. (1978) Feeding behaviour in a population of domestic fowls in the wild. *Applied Animal Ethology*, **4**, 13–27

Schjelderup-Ebbe, T. (1922) Beitrage zur Sozialpsychologie des Haushuhns. *Zeitschrift für Psychologie*, **88**, 225–252

Schmidt-Nielsen, K. (1975) Recent advances in avian respiration. In *Symposium of the Royal Society of London*, No. 35, Avian Physiology (Peaker, M. ed.), pp. 33–47. London, Academic Press

Seligman, M.E.P. (1975) *Helplessness: on Depression, Development and Death*. San Francisco, Freeman

Sewell, R.B.S. and Guha, B.S. (1931) Zoological remains. In *Mohenjo-Daro and the Indus Civilisation II* (Marshall, J. ed.), pp. 649–673. London, Arthur Probsthain

Shanawany, M.M. (1982) The effect of ahemeral light and dark cycles on the performance of laying hens – a review. *World's Poultry Science Journal*, **38**, 120–126

Sharp, P.J., Macnamee, M.C., Sterling, R.J., Lea, R.W. and Pedersen, H.C. (1988) Relationship between prolactin, LH and broody behaviour in bantam hens. *Journal of Endocrinology*, **118**, 279–286

Siegel, H.S. and Siegel, P.B. (1961) The relationship of social competition with endocrine weights and activity in male chickens. *Animal Behaviour*, **9**, 151–158

Siegel, P.B. (1965) Genetics of behavior: selection for mating ability in chickens. *Genetics*, **52**, 1269–1277

Siegel, P.B. and Beane, W.L. (1963) Semen characteristics of chickens maintained in all-male flocks and in individual cages. *Poultry Science*, **42**, 1028–1030

Siegel, P.B. and Siegel, H.S. (1964) Rearing methods and subsequent sexual behaviour of male chickens. *Animal Behaviour*, **12**, 270–271

Smith, W.K. (1981) Poultry housing problems in the tropics and subtropics. In *Environmental Aspects of Housing for Animal Production* (Clark, J.A. ed.), pp. 235–258. London, Butterworths

Smith, W.K. and Dun, P. (1983) 'What Type of Nest?' Unpublished Paper presented to British Poultry Breeders and Hatcheries Association Flock Farmers' Conference

Social Surveys (1968) *Factory Farming – What the Farmers really Think*. Survey carried out for the Coordinating Committee on Factory Farming. Newport, Isle of Wight, Social Surveys (Gallup Poll) Ltd

Soller, M. and Rappaport, S. (1971) The correlation between growth rate and male fertility and some observations on selecting for male fertility in broiler stocks. *Poultry Science*, **50**, 248–256

Sorensen, P. (1989) Broiler selection and welfare. In *Proceedings of the Third European Symposium on Poultry Welfare* (Faure, J.M. and Mills, A.D. eds), pp. 45–58. Tours, World's Poultry Science Association

Sykes, A.H. (1988) Intake of sodium following sodium deficiency in the laying hen. *British Poultry Science*, **29**, 884–885

Tauson, R. (1984) Effects of a perch in conventional cages for laying hens. *Acta Agriculturae Scandinavica*, **34**, 193–209

Tauson, R. (1985) Mortality in laying hens caused by differences in cage design. *Acta Agriculturae Scandinavica*, **35**, 165–174

Tauson, R. (1986) Avoiding excessive growth of claws in caged laying hens. *Acta Agriculturae Scandinavica*, **36**, 95–106

Tauson, R. (1988) Effects of redesign. In *Cages for the Future*, pp. 42–69. Cambridge Poultry Conference, Agricultural Development and Advisory Service

Tauson, R. (1989) Cages for laying hens: yesterday and today . . . tomorrow? In *Proceedings, Third European Symposium on Poultry Welfare* (Faure, J.M. and Mills, A.D. eds), pp. 165–181. Tours, World's Poultry Science Association

Tauson, R., Jansson, L. and Elwinger, K. (1991) Whole grain/crushed peas and a concentrate in mechanised choice feeding for caged laying hens. *Acta Agriculturae Scandinavica*, **41**, 75–83

Temple, W., Foster, T.M. and O'Donnell, C.S. (1984) Behavioural estimates of auditory thresholds in hens. *British Poultry Science*, **25**, 487–493

Thomson, A.L. (1964) *A New Dictionary of Birds*. London, Nelson

Thorpe, W. H. (1951) The definition of terms used in animal behaviour studies. *Bulletin of Animal Behaviour*, **9**, 34–40

Toates, F. (1986) *Motivational Systems*. Cambridge, Cambridge University Press

Toates, F. and Jensen, P. (1990) Ethological and psychological models of motivation – towards a synthesis. In *Simulation of Adaptive Behaviour* (Meyer, J-A. and Wilson, S. eds). Massachusetts, MIT Press/Bradford Books

Todd, P. (1989) The Protection of Animals Acts 1911–1964. In *Animal Welfare and the Law* (Blackman, D.E., Humphries, P.N. and Todd, P. eds). Cambridge, Cambridge University Press

Tolman, C.W. and Wilson, G.F. (1965) Social feeding in domestic chicks. *Animal Behaviour*, 13, 134–142

Tucker, S. (1989) Alternatives? *ADAS Poultry Journal*, **3(1)**, 15–30

Upp, W. (1928) Preferential mating of fowls. *Poultry Science,* **35**, 969–976

van Enckevort, J.W.F. (1965) Het werkgebied bij leghennen. *Veeteelt en Zuivelberichten* 530–536

van Kampen, M. (1981) Thermal influences on poultry. In *Environmental Aspects of Housing for Animal Production* (Clark, J.A. ed.), pp. 131–147. London, Butterworths

van Rooijen, J. (1985) Du-evidenz, applied ethology and animal welfare. *British Veterinary Journal*, **141**, 245–248

van Rooijen, J. (1989) *De Kip Alsproefkonijn in het Gedragsonderzoek*. Spelderholt, Centrum voor Onderzoek en Voorlichting voor de Pluimveehouderij

Vestergaard, K. (1980) The regulation of dust-bathing and other behaviour patterns in the laying hen: a Lorenzian approach. In *The Laying Hen and its Environment* (Moss, R. ed.), pp. 101–113. The Hague, Martinus Nijhoff

Vestergaard, K. (1982) The significance of dust bathing for the wellbeing of the domestic hen. *Tierhaltung*, **13**, 109–118

Webster, A.J.F. and Nicol, C.J. (1988) The case for welfare. In *Cages for the Future*, pp. 11–21. Cambridge Poultry Conference, Agricultural Development and Advisory Service

Wegner, R.M. (1981) Choice of production systems for egg layers. In *Proceedings, First European Symposium on Poultry Welfare* (Sorensen, L.Y. ed.), pp. 141–148. Copenhagen World's Poultry Science Association

Wegner, R.M. (1986) Alternative Systeme fur Legehennen-Untersuchungen in Europa. In *Proceedings, 7th European Poultry Conference* (Larbier, M. ed.), pp. 1064–1076. Paris, World's Poultry Science Association

Wegner, R.M. (1990) Experience with the get-away cage system. *World's Poultry Science Journal*, **46**, 41–47

Wennrich, G. (1975) Studien zum Verhalten verschiedener Hybrid-Herkunfte von Haushuhnen (*Gallus domesticus*) in Bodenintensivhaltung mit besonderer Beruchsichtigung aggressiven Verhaltens sowie des Federpickens und des Kannibalismus. 5. Mitteilung: *Verhaltensweisen des Federpickens. Archiv fur Geflugelkunde*, **39**, 37–44

West, B. and Zhou, B-X. (1989) Did chickens go north? New evidence for domestication. *World's Poultry Science Journal*, **45**, 205–218

Woodard, A.E. and Abplanalp, H. (1967) Semen production and fertilising capacity of semen from Broad Breasted Bronze turkeys maintained in cages and on the floor. *Poultry Science*, **46**, 823–826

Woodard, A.E. and Wilson, W.O. (1970) Behavioral patterns associated with oviposition in Japanese quail and chickens. *Journal of Interdisciplinary Cycle Research*, **1**, 173–180

Wood-Gush, D.G.M. (1959a) A history of the domestic fowl from antiquity to the 19th century. *Poultry Science*, **38**, 321–326

Wood-Gush, D.G.M. (1959b) Time-lapse photography: a technique for studying diurnal rhythms. *Physiological Zoology*, **32**, 272–283

Wood-Gush, D.G.M. (1960) A study of sex drive of two strains of cockerels through three generations. *Animal Behaviour*, **8**, 43–53

Wood-Gush, D.G.M. (1963a) The relationship between hormonally-induced sexual behaviour in male chicks and their adult sexual behaviour. *Animal Behaviour*, **11**, 400–402

Wood-Gush, D.G.M. (1963b) The control of the nesting behaviour of the domestic hen. I. The role of the oviduct. *Animal Behaviour*, **11**, 293–299

Wood-Gush, D.G.M. (1971) *The Behaviour of the Domestic Fowl*. London, Heinemann

Wood-Gush, D.G.M. (1972) Strain differences in response to sub-optimal stimuli in the fowl. *Animal Behaviour*, **20**, 72–76

Wood-Gush, D.G.M. (1973) Animal welfare in modern agriculture. *British Veterinary Journal*, **129**, 167–174

Wood-Gush, D.G.M. and Gilbert, A.B. (1964) The control of the nesting behaviour of the domestic hen. II. The role of the ovary. *Animal Behaviour*, **12**, 451–453

Wood-Gush, D.G.M. and Gilbert, A.B. (1970) The rate of egg loss through internal laying. *British Poultry Science*, **11**, 161–163

Wood-Gush, D.G.M. and Gilbert, A.B. (1973) Some hormones involved in the nesting behaviour of hens. *Animal Behaviour*, **21**, 98–103

Wood-Gush, D.G.M. and Gilbert, A.B. (1975) The physiological basis of a behaviour pattern in the domestic hen. *Symposia of the Zoological Society of London*, **35**, 261–276

Wood-Gush, D.G.M. and Kare, M.R. (1966) The behaviour of calcium-deficient chickens. *British Poultry Science*, **7**, 285–290

Wood-Gush, D.G.M. and Murphy, L.B. (1970) Some factors affecting the choice of nests by the hen. *British Poultry Science*, **11**, 415–417

Wood-Gush, D.G.M. and Osborne, R. (1956) A study of differences in the sex drive of the domestic cock. *Animal Behaviour*, **6**, 68–71

Wood-Gush, D.G.M., Duncan, I.J.H. and Savory, C.J. (1978) Observations on the social behaviour of domestic fowl in the wild. *Biology of Behaviour*, **3**, 193–205

Yamada, Y. (1988) The contribution of poultry science to society. *World's Poultry Science Journal*, **44**, 172–178

Yeomans, M.R. (1987) 'Control of Drinking in Domestic Fowls'. PhD thesis, University of Edinburgh.

Zeuner, E.E. (1963) *A History of Domesticated Animals*. London, Hutchinson

INDEX